Waiting for Macedonia

Waiting for Macedonia
Identity in a Changing World

Ilká Thiessen

BROADVIEW ETHNOGRAPHIES & CASE STUDIES

broadview press

Library and Archives Canada Cataloguing in Publication

Thiessen, Ilká

Waiting for Macedonia : identity in a changing world / Ilká Thiessen.

(Broadview ethnographies & case studies)
Includes bibliographical references and index.
ISBN-13: 978-1-55111-719-5
ISBN-10: 1-55111-719-3

1. Women engineers—Macedonia (Republic)—Skopje—Social conditions. 2. Women engineers—Macedonia (Republic)—Skopje—Attitudes. 3. Social change—Macedonia (Republic). 4. Post-communism—Macedonia (Republic). 5. Macedonia (Republic)—Politics and government—1945-1992. 6. Macedonia (Republic)—Politics and government—1992-. I. Title. II. Series.

HN640.S54T48 2006 305.43'620094976 C2006-904419-8

Broadview Press is an independent, international publishing house, incorporated in 1985. Broadview believes in shared ownership, both with its employees and with the general public; since the year 2000 Broadview shares have traded publicly on the Toronto Venture Exchange under the symbol BDP.

We welcome comments and suggestions regarding any aspect of our publications—please feel free to contact us at the addresses below or at broadview@broadviewpress.com.

North America
PO Box 1243, Peterborough, Ontario, Canada K9J 7H5
PO Box 1015, 3576 California Road, Orchard Park, NY, USA 14127
Tel: (705) 743-8990; Fax: (705) 743-8353
email: customerservice@broadviewpress.com

UK, Ireland, and continental Europe
NBN International, Estover Road, Plymouth, UK PL6 7PY
Tel: 44 (0) 1752 202300; Fax: 44 (0) 1752 202330
email: enquiries@nbninternational.com

Australia and New Zealand
UNIREPS, University of New South Wales
Sydney, NSW, Australia 2052
Tel: 61 2 9664 0999; Fax: 61 2 9664 5420
email: info.press@unsw.edu.au

www.broadviewpress.com

Consulting Editor for Philosophy: John Burbidge

Printed in Canada

Contents

List of Tables and Figures

TABLES

FIGURES

Acknowledgements

My first debt of gratitude is to the many people in Macedonia who helped me carry out this research. Words of gratitude should be repaid with open thanks, yet confidentiality requires anonymity. It is impossible to express adequately my debt to my friends in Macedonia. I shall, however, name some people who, although not a direct part of my research, have been more than generous to me with their help and friendship: foremost are Anita Dimova, her brother Igor Poposki, and her parents Maria and Kiro Poposki. They shared their families and friends with me and provided a true home, a place of comfort and reflection. And I would like to thank Tamara Dimova for the hope and future she brings. My deepest thanks also go to Nikica Kusinikova and her family. It is nearly impossible to express the depth of my gratitude to all of the people involved in this book. My most wonderful experience of the research from which this book originates was the time I spent with all of you! I thank you for your kindness and deep friendship.

I also wish to thank Simon (Sime) Bojadzievski and the late Nikola (Koki) Smilevski, who have always been there for me in times of joy and pain; Shaip, Mimi, and David Jashari, whom I would like to thank for their extraordinary generosity and friendship. I would also like to express my appreciation to the staff of the German embassy who have been helpful throughout my time in Macedonia.

My supervisors at the London School of Economics (LSE), Peter Loizos and Fenella Cannell, I have to thank, not only for their encouragement, advice, challenge, and trenchant criticism but also for their great tolerance and patience. Bernhard Streck from the University of Mainz and Leipzig has shared his expertise and given me encouragement to start this research, and I am greatly indebted to him. Frances Pine from Cambridge University looked in on my writing at various times and gave some very helpful comments particularly for Chapter Four and introduced me to some valuable literature. I have also benefited from discussions of my material with Keith Brown and David Rheubottom. First versions of Chapter Two were presented in December 1995 at the Anthropological Understandings of the Politics of Identity Seminar, Queen's University, Belfast; in May 1996 at the London School of Economics Greece and the Balkans Seminar; and in July 1996 at the EASA (European Association of Social Anthropologists) conference in Barcelona (in the Town and Country in North Africa and South Europe Workshop organised by John Davis and Maria Angels Roqué and funded by the London School of Economics and Political Science Graduate Research Fund). I wish to thank all the participants at all conferences and seminars who shared their comments with me. I also would like to thank my two examiners Frances Pine from Cambridge University and Wendy Bracewell from the School of Slavonic and Eastern European Studies for their helpful comments.

My field research in Macedonia from 1994 to 1995 was made possible by the Deutscher Akademischer Austauschdienst e.V. (DAAD). I was further fully funded by the Studienstiftung des Deutschen Volkes from 1995 to 1999. I would like to express my great gratitude to the Studienstiftung, which has not only financed my studies at LSE, but also given me extraordinary, challenging, and enthusiastic support since 1986.

I would like to thank Alison Renouf so much for her friendship and her incredible and patient help in correcting my first manuscript. At LSE, my thanks go to Todd Sanders and the late Bwire Kaare for their friendship and support that helped me tremendously in overcoming "post-research trauma." I would like to thank Sini Cedercreutz who has been a constant confidante. I also would like to thank Gillian and Dr. David Bird for their kind support and Christine Schollmayer for her friendship.

My greatest gratitude goes to my parents, Gislinde and Jörn Thiessen, and to my husband, Clinton Jay Kuzio; all three have shared the most painful times of this thesis and have always stood by my side. I would like to dedicate this book to them and to my aunt, the late Ingelore Weiche, whose curiosity and willpower have been inspirational.

Living in Macedonia has been a privilege. I hope that despite all the mistakes and faults of this book, it is recognized as written with the deepest gratitude to all the people of Macedonia.

Introduction

Western media continue to portray Eastern Europe as a troubled place, rife with ancient hatreds, tribal affiliations, and methods of mass terror supposedly abandoned in the modern World. The West, long eager for a peek behind the Iron Curtain, welcomed Eastern Europe into the "free world" in 1989 with the fall of the Berlin Wall. High expectations accompanied this welcome: it was obvious that the Cold War finally had ended and that, from now on, the world could turn into a better place.

I am intrigued by the continuous ignorance of the so-called free world—the capitalist world, the world of democracy and human rights—concerning what has happened in Eastern Europe and even about what Eastern Europe actually is. In the late 1980s when Eastern Europe came tumbling towards the "free world" (and I am using this expression with a sense of irony), nobody seemed to have really thought about what this would mean in the reality of daily life. What was Eastern Europe? Is Eastern Europe a political entity? Does Eastern Europe have its own identity? The fall of the Berlin Wall caught all by surprise, myself included; listening to the news I indeed cried from happiness. Today I cherish 1989 and 1990. It was an exciting time to live in Europe. It was a time when I had the privilege to start the research from which this book originates. I distinctly remember the spark, the joie de vivre that pulsed through Eastern Europe at this time. In Yugoslavia, Ante Markovich had done an extraordinary deed and managed to stabilise the dinar, the Yugoslav currency, which mounted to an exchange rate of one to one with the German Mark. It still rings in my ears, the excitement of the people I met in Yugoslavia during this time: "Soon we will be part of the EU—we will be Europe." Then the war in Bosnia began, and Europe became divided into "Western Europe," that is the European Union, and Eastern Europe.

Eastern Europe in the Western media became portrayed as a dark place that had to be led into the modern age. However, Europe and the Western World were overlooking something vital in their disdain for Eastern Europe and its "backwardness," which, to the West, seemed to be the origin of the wars in the Former Yugoslavia. I insist that Eastern Europe as it is today was created, not by communism but by the so-called free world. I know this, too, is a simplified argument, but I believe it is an argument that must be made in order to understand the complexity of any political analysis that centres on the "identity" of a society.

As an example of this complexity, consider that most studies done on Eastern Europe during the Cold War looked at economics and development, state farms and collectivisation. The question is why researchers decided to concentrate on these specific aspects of Eastern Europe. In my work, I look at normal day to day life in Macedonia using common anthropological methods to understand the macrocosm through the microcosm and not the other way around.

My first question to the reader is, do we really live in a world that is determined by post-socialism? I would argue not. If not, what can we conclude? Is it a Western misconception that socialism is over, is post, is gone and done with? I believe stressing the so-called demise of socialism would be an over-simplification and would not lead us to an understanding of the political reality of Eastern Europe. In societies that go through dramatic social changes, each and every individual has to define herself or himself anew. People experiencing these changes have to reflect on their lives in a far more complex manner than, say, the average Canadian citizen would. To identify this process of reflection and social change as "post-socialism" denies the complexity of the issue. Every day individuals in society have to make new decisions about what is right and what is wrong, and making these decisions is problematic when daily life routines are becoming questioned, becoming continuously politicised. I do not need to point out that many Eastern European citizens soon discovered that liberal democracy is not necessarily always the right thing and that, on ideological grounds, some things that socialism brought were beneficial. But in many of the government reports—reports on economic development and on human rights, reports that are produced en masse by Western government agencies and NGOs—the complexity of people's mundane but politicised daily life is largely ignored.

Consequently, I want to offer three things with this book. First, my study of the lives of a group of young female engineers in Skopje, Macedonia between 1988 and 1996 draws a picture of what actually happened during this time in Europe and in Macedonia specifically. Second, this picture offers a way of understanding the ethnic conflict in Macedonia, a conflict that still has the potential of escalating into another civil war in the Balkans. I hope that such an understanding will critically reflect on the role that the West has played in the creation of such ethnic conflicts. Third, I offer a vision of what Macedonia "is" and "will be," taken directly out of my fieldwork experience. To assess the past and envision the future of this region, we have to first understand our own concepts and notions of Eastern Europe. From this understanding, I hope we can create policies appropriate to the concerns of the region, and that we can listen to its voices. Quite conceivably, the West itself can learn from Eastern European people's experiences with concepts such as democracy and free market. Eastern Europe could well offer the West very adaptable alternatives.

DEFINING THE SETTING

T'ga za jug (Longing for the South)

If I had an eagle's wings
I would rise and fly on them
To our shores, to our own parts,
To see Stamboul, to see Kukush;
And to watch the sunrise: is it
Dim there too as it is here?

If the sun still rises dimly,
If it meets me there as here,
I'll prepare for further travels,
I shall flee to other shores
Where the sunrise greets me brightly,
And the sky is sewn with stars.

It is dark here; dark surrounds me,
Dark fog covers all the earth,
Here are frosts and snows and ashes,
Blizzards and harsh winds abound.
Fogs all around, the earth is ice,
And in the breast are cold, dark thoughts.

No, I cannot stay here, no;
I cannot look upon these frosts.
Give me wings and I will don them;
I will fly to our own shores,
Go once more to our own places,
Go to Ohrid and to Struga.

There the sunrise warms the soul,
The sun sets bright in mountain woods:
Yonder gifts in great profusion
Richly spread by nature's power.
See the clear lake stretching white-
Or blue darkened by the wind,
Look at the plains or mountains:
Beauty everywhere divine.

To play the flute there to my heart's content!
Ah! Let the sun set, let me die.

(1861)
Konstantin Miladinov

Longing for the South. This poem is dearly remembered in Skopje and, not surprisingly, several cafés in the capital city of the independent Republic of Macedonia are named *T'ga za Jug*. The poem speaks of the yearning for a country that is far away and seems to promise all that is missing. Miladinov talks about beauty and darkness, about the wish to fly freely wherever the wind directs, about ultimate death. When I am asked to define Macedonia, *T'ga za Jug* springs to my mind.

It was in another café that *T'ga za Jug* was put to me as "Waiting for Macedonia" with a meaning similar to that of Beckett's *Waiting for Godot*. "Nothing to be done" is the starting sequence. Dialogue has no function. Macedonia as non-subject of action, from neither the inside nor the outside: the story of Macedonia neither creates nor resolves conflicts, does not develop either ethical or political programs of reform, and it does not offer meaning for its human existence. Every interpretation is null. There is no direction and no aim. This is how a great number of people from different backgrounds and in differing circumstances in Macedonia have described Macedonia to me over and over since its independence. Those I was privileged to live with during my research described people in Skopje as waiting for Macedonia, waiting for their country to come into being. It is this world I will write about, though I believe that Macedonia is "becoming." Nevertheless, I have to use this perception of Macedonia as my vantage point in order to understand what Macedonia was, is, and will be about for my group of friends.

It is a warm July weekend. Zenia and I are sitting in Café Ciao in the afternoon, waiting for our friend Boris to come. I am taking my notes, and Zenia is laughingly suggesting that I should call my research "Waiting for Boris" as we spend so much time doing just that. Then she looks at me seriously and says, "Ilka, maybe you should call your thesis 'Waiting for Macedonia.'" I did spend considerable time waiting *in* Macedonia, but, more importantly, I spent a lot of time waiting *with* my friends. *Waiting for Macedonia*. Later in the evening, a stranger in Café ZZ-Top speaks, "We are waiting, always waiting, all our lives waiting, as our parents and grandparents have. Our parents once thought they had found it, but now they have lost it, and we are waiting again." He was referring to a version of Macedonia's past, present, and future. Today, both children and parents believe that in socialist Yugoslavia, Macedonia found what it had been looking for—an identity. They had an identity as the socialist Republic of Macedonia, an integrated part of Yugoslavia complete with pride and freedom. With the disintegration of Yugoslavia, *this* identity vanished. However, Macedonia did retain its independence, a fact of no interest to Serbia at that time but of interest to the international community.

Macedonia was already seen as the explosive barrel in the Balkan conflict that could have involved Greece and Turkey in the early nineties. Macedonia declared independence in 1991, and it seemed easy enough to create an independent democratic Macedonia on the foundation of the socialist republic. Institutions, the legal and political apparatus, were retained; privatisation was initialised through great economic help from a wary Europe and an eager USA. Nevertheless, how could Macedonia be defined on a personal level? How did the citizens of the newly independent Macedonia deal with the changes around them? What did Macedonia mean for them, what did it mean to be Macedonian, and how could their identity be redefined?

ANTHROPOLOGY OF THE BALKANS

Much has been written about the economic and political changes in Eastern Europe, but a consideration of how these changes construct personal lives has been neglected.[1] Brailsford gives us the following information about Macedonia:

> I have heard the French consul declare that with a fund of a million francs he would undertake to make all Macedonia French. He would preach that the Macedonians are the descendants of the French crusaders who conquered Salonica in the twelfth century, and the francs would do the rest. (Brailsford [1906] 1971: 103)

This identity vacuum does not apply to today's Macedonia. However, the general perception of Macedonia, by Macedonians and the outside world alike, as something that has to be filled by others, prevails. But recent developments in Macedonia might very well show that Macedonia can be politically and socially viable on its own, despite recent conflict.

There have been some excellent books written about the region in the recent past: Bringa (1991, 1995) on Bosnia-Herzegovina, Karakasidou's work on Greek-Macedonia (1997a), and Keith Brown's multiple works on Macedonia. Despite this recent peak of anthropological interest, Yugoslavia or Macedonia otherwise seemed to have been under anthropological investigation only during the 1960s and 1970s, when rural adaptation, kinship, and tradition were under consideration.

So the "Republika Makedonija" presents an unusual and relatively unexplored field of research, in that it relates to the specific circumstances of Yugoslavia: Tito's socialism, the disintegration of Yugoslavia, and the war that has shattered the lives of so many in former Yugoslavia. For me, Macedonia represents its own specific case of the current rapid and profound changes occurring in a society that belonged to the former socialist Yugoslavian Republic. What are the defining parameters of Macedonia? First we have to recognize the achievement of nationalist integrity under Tito: In my interviews, Macedonians argued that they felt Albanians and Roma were Macedonians just as themselves while being in Yugoslavia. We then have to look at the challenge to this identity, imagined or real, experienced by the Republic of Macedonia after 1991, at the hands of its neighbours

Greece, Serbia, Bulgaria, and Albania. Then there are the animosities within Macedonia between Macedonians and Albanians, which have to be understood in relation to the fighting in Kosovo. Even today, there lingers the fear of a civil war between the Slavic-Orthodox Macedonians and the Albanian population, fuelled by their definition of difference. Then there is the constant reference to the historical inheritance of 500 years of Ottoman rule, and to Macedonia's strong patriarchal and agricultural past in the midst of its present unusually high level of urbanisation. I believe all these aspects render Macedonia a highly complex and dynamic site of social change.

In view of these circumstances, I will concentrate on the urban environment of Skopje. There have been studies about the urban environment of Yugoslavia, though the research has dealt exclusively with the adaptive character of traditional culture or has focused on the role of kinship, linking rural and urban sectors of society.[2] In Macedonia, Spangler points out that urban anthropological research is almost non-existent because the rural population of Macedonia has traditionally been viewed as more interesting (Spangler 1983: 91). In respect to the present situation of Macedonia, I wish to challenge this perspective and draw attention to the specific cultural environment of a new state in the city of Skopje.

Another argument, which is predominant in the literature about Yugoslavia, is one that leaves the impression that kinship is one of the most important features in Yugoslavia.[3] By considering the circumstances of young female engineers, I wish to challenge the classical perceptions of women as passive receivers of change as presented by Lockwood (1975), who writes about peasantry and the marketplace in Bosnia and about women solely in the context of marriage. I also wish to challenge scholarship that ignores women, such as that of Ford (1982) and Brown (1995), who do not discuss women at all. Boehm's *Blood Revenge: The Anthropology of Feuding in Montenegro and Other Tribal Societies* presents a good example of male-centred ethnography. Boehm seems to assume that in a world of feuding there is no place for women. Women again are seen only in their reproductive kinship roles. Women receive little attention from Rheubottom either, though in "The Seed of Evil Within" he offers a perspective of Macedonia that my group of graduates choose consciously or unconsciously to contradict. Women in these accounts do not seem to have a life of their own. How do they relate to the feuding described in Boehm's account? No answer. The spaces occupied by women in society other than in the realm of kinship are overlooked. Although published in 1984, Boehm's research dates from 1963 to 1966. In his and in Rheubottom's and Lockwood's accounts, the element of possible change in this region is almost not worth mentioning.

I am setting out to challenge these works by looking in particular at the life of young professional women in Skopje and by discussing current change as the dominant aspect of social life in Macedonia through a gendered perspective investigating the intersection of gender and national identity. I am specifically looking at the "masculinising effect" of liberal democracy. Although the patriarchal notions of the new nationalism arising in the region are accounted to the backwardness of Eastern Europe, it should not be forgotten that socialism had once set out to erase gender differences and that the inheritance of this aspect of socialist ideology seems to have been swept away by liberal democracy. It is social change that gives new ways of talking about society and about individuals' experiences in society. It is the situation of profound change that offers new means of creating and interpreting social order and, beyond that, new ways of ordering and talking about life itself.

WHY FEMALE ENGINEERS?

I concentrated on young, urban, educated women and their lives and on their relations with men, parents, and grandparents, considering these women during different phases and amidst various circumstances—love, marriage, education, work, and social life—in order to give broader insights about how social change is perceived and perpetuated. In such a setting, a gender-based analysis of political and economical restructuring will provide a deeper view of how this change is configured and points to future developments.

The history of socialism in Eastern Europe originated from a revolutionary change of social relations, including gender relations. In the aftermath of socialism, the question remains as to the direction in which the new states will venture. In many former socialist countries (i.e., Poland and Croatia), the resurgence of a Catholic inheritance depicts the image of the woman as "The Mother." Feminist programs have been sacrificed to the cause of progress and liberation. Women have been "reassigned" to their former domestic roles. To look at social change through a gendered perspective, then, can illuminate and help us evaluate the very processes of social change. Women's status in society is not static: it is both influenced by and has an effect upon the process of social transformation itself (Einhorn 1993: 2–3). Despite some pessimistic undertones in scholarship on the current situation of women in Eastern Europe (e.g., Phizacklea, Pilkington and Rai 1992), I believe Macedonia could offer a different picture. Although the Republic of Macedonia is in desperate need of a new state ideology, which separates it from any claims that could be made on its sovereignty by Serbia, Bulgaria, Greece or Albania, it does not choose to condemn the socialist era entirely, nor its inheritance of social reform. It was Tito who gave Macedonia nationhood by forming the People's Republic of Macedonia. This aspect, as well as the nationalist-inspired war in former Yugoslavia, gives Macedonia a very different outlook on its own definition and on any re-definition of women's roles in society.

It is younger women who have to be studied, along with their reflections on their parents and grandparents, to understand future developments of change. Certain aspects have to come under close review. A high proportion of female university students was common in all former state socialist countries. Despite this, there seemed to be a tendency for them to look for training in areas deemed "suitable for women" (Einhorn 1993). I, therefore, chose as my informants young women enrolled in the Faculty of Electrical Engineering at the University of Skopje, which has a high proportion of female students despite the electrical engineering field's traditional male dominance. I observed their life and discovered how they make their way into employment and in their social and family lives. I also considered whether there is a search for new political and social values, and, if so, how these are constructed and what relationship they have with the past. To understand the conflicts and the environment of change in Eastern Europe, one has to validate women's subjective contradictory experience of continuity and change.

My research setting is the urban environment of Skopje, the Faculty of Electrical Engineering, and the surroundings experienced by a group of friends. It is this group that seems to be extremely vulnerable to the current change in Macedonia. By connecting these women and men to the dramatic and profound changes in their lives as they experience the loss of one national identity (Yugoslavian) and the creation of a new one (specifically Macedonian), I hope to understand how they construct their personal and gendered identity within this specific setting. There is much evidence to indicate that this specific group—young, urban, and highly educated—contains the early innovators. They are motivated to distinguish themselves from the rest of the population and create a cycle of innovation and diffusion. Social change is and was a dominant force in the last forty years of Macedonian history. Macedonia moved from being an agricultural society to a modern, industrial society, and, today, social change again might lead Macedonia into true independence.

A NOTE ABOUT METHODOLOGY

This book is based on research conducted between 1988 and 1996 in Macedonia, but my interest in Macedonia is continuous and cannot be put into a time-frame. It was the people who drew me to Macedonia in the first place. I had the incredible privilege of getting to know some very special people. Consequently, the reader will note that I often talk about my friends rather than my informants. This designation is indicative of the personal relationships I had as well as of my academic position that the people we write about are not objects of our investigation but subjects of a co-operation. Despite this personal engagement, which is, in my opinion, academically necessary to address the microcosm of daily life in any meaningful way, the analysis of this personal data is deeply rooted in the methodology of Social Anthropology. I used Participant Observation: more than one hundred two- to three-hour open-ended interviews in Macedonian, which have been professionally translated; several directed interviews; and about ten life histories. Most interviews I conducted alone, although some I did with a translator. Occasionally, interviews had a family member present. These interviews took place either in the home of the interviewees or in my own home. I tape-recorded most interviews, though sometimes, when asked, I would simply make notes.

A NOTE ABOUT THE HISTORY OF MACEDONIA

Macedonia is a complex site, in the political as in the historical sense. In looking at Macedonian history, I had the choice of looking at the Macedonian, Greek, Bulgarian, or Serbian interpretation of historical events, each with its own political agenda. Even a timeline is not politically neutral. With which event do we start? Which historical moments do we stress? At this point I would like to leave the presentation of Macedonia's history to the historian, Macedonian, Greek, Serbian, Bulgarian, or foreign. Writing Macedonia's history is complex and contested and merits a book on historiography and politics.

Macedonia has invoked some political and economic interest from Western countries, not because of its rich history but because of its strategic location across a major junction, the north-south route from the Danube River to the Aegean formed by the valleys of the Morava and Vardar rivers and the ancient east-west trade routes connecting the Black Sea and Istanbul with the Adriatic Sea. Today, most Macedonians in the Republic do not claim any connection with Alexander the Great but see themselves as Slavs. However, the region has been under the rule of the Ottoman Empire for more than 500 years and has incorporated a substantial number of other ethnic groups, including Albanians, Vlachs, and Roma.

In Chapter One, I explore the Macedonian context and identify important political issues. Chapter Two looks into the mapping of urban identity and connects specific themes that have been introduced in the previous chapter. I make these connections by exploring the landscape of the city of Skopje and its role in shaping a discourse on identity for my informants. The city comes to stand in opposition to the village and within the city the north and the south come to stand for the difference between Albanian and Slavic Macedonians. The construction of Skopje becomes a discourse of the past with the future. Chapter Three looks at how differences are created in Macedonia and theorises about their effects. Chapter Four deals with the changes in Republika Makedonija that become visible when looking at how the female graduates enter the world of employment under the current circumstances. I explore what is changing and what persists in the Republic of Macedonia. Within this framework, I discuss social and physical mobility and status and lifestyle expectations. In Chapter Five, consumption is presented as a communication with the world outside Macedonia. In Chapter Six, I discuss the symbolic inscription of the bodies of my informants. Their body management represents security in a world of uncertainties.

NOTES

1. Extant scholarship is often historically oriented (i.e., Halpern 1972) or deals with specific cultural traditions (Rheubottom 1971). More prominently, studies have been conducted in, for example, Hungary (Stewart 1987), Poland (Pine 1987), and Romania (Kligman 1988). Several anthropologists have worked in Yugoslavia in the context of the anthropology of kinship, primarily, although not exclusively, on the zadruga system (for the zadruga see Baric' 1967a; Denich 1974; Halpern and Anderson 1970; Hammel 1972; Mosely 1976; Rheubottom 1980; Erlich 1966. See also the work *Communal Families in the Balkans: The Zadruga,* ed. Robert F. Byrnes [Notre Dame, IN: University of Notre Dame Press, 1976]). In social anthropology, Macedonia itself has been studied by Rheubottom (1976a, 1976b, 1980, and [1985] 1993); Ford (1982), who deals with the adaptive character of traditional culture; Brown (1995), who looks at the issue of nationalism; and Brailsford, whose work dates back to 1906.
2. See Baric' 1967a, 1967b; Halpern 1963; Hammel 1969; Hammel and Yarbrough 1973; Simic' 1973a, 1973b, 1974.
3. See, for example, Baric' 1967b: 266 and Rheubottom 1980: 222.

Waiting for Macedonia

Chapter One

Macedonian Context

Who can be called Macedonian? Is there a Macedonian past and history? Is there a Macedonian identity that the people of today's Republic of Macedonia can claim as theirs? *Who Are the Macedonians?* (Poulton 1995) and *The Macedonian Conflict* (Danforth 1995) are intriguing book titles that point toward the dilemma within the Republic of Macedonia today. There is no doubt that Macedonia did, and does today, exist in a geographical and historical sense, although, with the fall of Yugoslavia, Macedonia's geography and history have been called into question. This situation, of suddenly being unable to describe or know who you are and what your people are, results in a void I call a vacuum. "*Who are you, if other people tell you that you do not exist?*" I heard this question many times during 1991 and 1995 while I was in Macedonia.

The path to independence for the Republic of Macedonia was direct. A popular referendum was held, and Macedonia seceded from Yugoslavia, becoming a sovereign state on September 8, 1991 when the majority of voters chose independence. The constitution was declared on November 17, 1991. On December 19, 1991, Macedonia was declared an independent state. During the following month, the Republic of Macedonia was recognised by about a dozen states. However, not until April 8, 1993 was Macedonia accepted within the United Nations as Macedonia, and, bowing to Greek pressures, the European Community then accepted the state only as *de facto.* The Greek hesitancy to acknowledge the Republic of Macedonia was based on its calling itself Macedonia, which Greece saw as the Republic exercising territorial claims upon the Greek Northern regions, arguing that the true Macedonians have been of Greek nationality since 2000 BC. Greece alleged that Macedonia had no right to call itself "Macedonia" because Macedonia has always been, and still is, a region of Greece.

As a member state of the European Community and a NATO military base, Greece was not to be opposed publicly; the EU never openly declared Greece's fears unwarranted, and they were because Macedonia was too small and too dependent on the international community to be of any serious threat to Greece.

In 1993, two years after becoming a sovereign state, Macedonia was viable as an independent state, despite economic and political difficulties and the pervasive fear of being pulled into the Yugoslavian civil war. Macedonia was protected from this pull through international forces, in particular through those of the US and UNPROFOR (United Nations Protection Force).

In 1993 and 1994, Macedonia had a highly plural government, as did other post-communist eastern European countries during this time; 40 parties soon formed 3 government blocks. The dominant political force is the middle-left, a union between social democrats, socialists, and liberals. Other parties, the rightist Macedonian and Albanian nationalist block and the Macedonian nationalists, VMRO, had dwindling public support during this time, in part because they lacked a political and economic program for Macedonia. A general poll from 1993–94 illustrates the trend toward a social-democratic system during this time.[1] Today, after the latest election held November 14, 1999, the current cabinet is formed by the following government coalition parties: VMRO-DPMNE (Internal Macedonian Revolutionary Organization-Democratic Party for Macedonian National Unity), LP (Liberal Party), and DPA (Democratic Party of Albanians).

Economically, Macedonia faces a mild crisis in that its links and routes to outside countries, and to their food and imports, for instance, are severely restricted due to the conflicts in the region and Macedonia's landlocked position. The international community, however, which has a strong interest in keeping the region peaceful, supports Macedonia in its slow and moderate economic change. It is feared that an ethnic conflict in Macedonia could spread and involve Turkey and Greece in the long run. To avoid conflict with Greece, Macedonia stresses, at least officially, that Macedonia's legitimisation as an independent state is based not on the glorious past of Alexander the Great but on the last fifty years as part of the former Yugoslavia.

However, today Macedonia's large Albanian minority and the *de facto* independence of neighbouring Kosovo have become greater and greater sources of ethnic tension within Macedonia.

YUGOSLAVIA AND MACEDONIA

During World War II, the Anti-Fascist Council for the National Liberation of Yugoslavia (AVNOJ) endorsed a Macedonian nation in 1943 and gave it equal status with the other five federal units of the future Yugoslavia. The Stalin-Tito split marked the end of any Yugoslav-Bulgarian co-operation and the end of any plans to unite Vardar, the Yugoslavian part of Macedonia, and Pirin Macedonia, the Bulgarian section. It was under Tito that a distinct Macedonian consciousness developed and during his rule when the terms "Macedonia" and "Macedonian" were used to represent an identity and to convey concepts that had no connection with Greece.

MACEDONIA AND "OTHERS"

Macedonia differs from other former republics and post-communist states because it has kept many of its social laws while changing its economy to a market economy and its politics to democracy in a more tender move away from socialism than many other Eastern European countries have made. Macedonia seeks stability in inter-ethnic relations in order to prevent itself from suffering a fate similar to that of other ex-republics of the former Yugoslavia. However, today these efforts seem to break down, exacerbated by the Kosovo conflict. Because Macedonia adheres to CSCE (Commission on Security and Cooperation in Europe) standards and is monitored by many human rights organisations, it is in a position to be recognised by the EU. Since 1991, the international community has guaranteed Macedonia's external security and co-ordinates internal reforms and Macedonia's economic transformation. Macedonia has pursued a policy of disarmament, conflict avoidance, and international co-operation.

In 1994, the ruling coalition in Macedonia won enough support to amend the constitution. Most importantly, the coalition changed annexing laws, which could be interpreted as encouraging the annexation of external territory. This paragraph of the constitution had been drafted in the early years of independence without much foresight, likely with the intention of rejoining Yugoslavia. Nevertheless, Greece had, with justification, protested loudly against this constitutional amendment, especially in consideration of Macedonian expatriates (Greece's Slav minority who had lived in Greek [Aegean]-Macedonia) wanting to return to Greece (see Danforth 1995).

These expatriates were criticised by my informants for having influenced the discussion about Alexander the Great being the "Ancestor of all Macedonians." It was these expatriates who were seen as the fervent nationalists appropriating Alexander the Great's symbols. The rhetoric of the expatriates was picked up by Macedonian nationalists within the Republic but was not well supported for two main reasons. First, Macedonians see themselves as descendants of the Slav migration who, by the mid-sixth century, had begun to settle into the Balkans and Macedonia. Second, Macedonians within the Republic of Macedonia felt that such rhetoric would put them in grave danger of repercussions, as it could quickly lead to war. Returning émigrés with their romanticised understanding of Macedonia's history have made little contribution to Macedonia's successful "non-extremist" policies and are met with much suspicion in Macedonia.

The Albanians form the largest national minority in Macedonia, and, because of their numbers, they claim official status as Macedonia's second people. They boycotted the independence referendum and census as they felt they did not receive enough recognition. Already in 1992, when the Macedonian-Greek conflict was at its height, there was the perception in Macedonia that the Albanian minority constituted a threat that could result in civil war.

Figure 1.1
Albanian and Macedonian Representations

INDEPENDENCE

The Greek-Macedonia conflict arose on September 9, 1991, when Macedonia held a referendum on its sovereignty and independence and on November 17 of the same year, when a new Constitution was adopted. By April 8, 1993, the Republic of Macedonia became a member of the United Nations. On September 13, 1995, Greece and the Republic of Macedonia signed an agreement regarding the Star of Vergina, a symbol found on the tomb of Philip II, father of Alexander the Great, which had been included on the Macedonian flag. Macedonia agreed to remove this potent symbol, and Greece responded by lifting the economic embargo that had severely disrupted Macedonia's economy. Further, it was agreed that negotiations over the issue of Macedonia's name would take place in the future, as Greece felt it had exclusive rights to the name and to Alexander the Great's historical presence. But in this highly politicised discussion of Macedonian history, a discussion that occurred during my fieldwork, several vital questions arose for many Macedonians: who are we as a nation, why are we a nation, and what does it mean to be Macedonian? Questions whose answers today seem to threaten the existence of Macedonia and, because of escalating ethnic tensions, to throw Macedonia into the next ethnic war in the Balkans.

WHO ARE THE MACEDONIANS?

Macedonians as a people consider themselves Slavic and Christian Orthodox. Greece, Bulgaria, Serbia, and Albania understand Macedonia as a particular geographical region and not a nation: they do not believe the people of that area have a specific "Macedonian" identity. In the Republic of Macedonia, a large minority of Muslim Albanians refuse to identify with the Macedonian state because of their differing religion and culture. They entertain the notion that West Macedonia will separate from the Republic and join Albania, unrealistic as this idea may be. As a state, the Republic of Macedonia currently considers people who have lived within the boundaries of the former Yugoslav Republic of Macedonia for the last 15 years to be citizens. The identity of Macedonia as a nation, however, is contested and there are overlaps between the self-definitions of Macedonia as a state and Macedonians as a people. This contention arises not through external competition but through internal conflict between the Slavic-Macedonians and the Albanian population.

From my perspective, within the boundaries of the Republic of Macedonia, there are Albanian, Vlach, Rom, Serbian, and Macedonian citizens, all with equal constitutional rights. Because of this, I make a distinction between citizens of Macedonia and Macedonians. I consider Slavic-Macedonians to have an ethnic identity, as there exists amongst them a definite sense of themselves as Macedonian. This sense of group identity has been acquired over the past 50 years while Macedonia was the Socialist Republic of Macedonia within the Yugoslav federation, a fact that, in my opinion, has been neglected by Macedonians, politicians, and academics alike.

GEOGRAPHICAL MACEDONIA

Macedonia is a term known for the last 3000 years, but the etymology of the country's name is still unclear. During the 500 years of Ottoman Rule, Macedonia, at that time an area of 67,500 km² with about 3.5 million people mostly of Slav origin was considered an important part of the Ottoman Empire. In the Balkan wars of 1912–13, Macedonia was split amongst Greece, which acquired over half the territory called Aegean-Macedonia; Serbia, which acquired the economically viable Vardar-Macedonia in which most of the Macedonian population lived; and Bulgaria, which gained a small area called Pirin-Macedonia. After World War I, Albania gained some Macedonian territory as well, specifically the west shore of the Ohrid Lake, with a population of approximately 100,000 Macedonians.

Figure 1.2
Map of Macedonia

Today's Republic of Macedonia is Vardar-Macedonia, the area Serbia received in 1913 as South Serbia and which it called the Socialist Republic of Macedonia. The republic has an area of 25,713 km^2, with 2,033,964 inhabitants, who declared their ethnic identity in 1991[2] as follows:

Table 1
Declared Ethnic Identity, Socialist Republic of Macedonia, 1991

Macedonian	65.3%
Albanian	21.7%
Turkish	3.8%
Rom	2.5%
Vlach	0.3%
Others	6.4%

In the last 50 years, Macedonia has experienced rapid urbanisation, and the Macedonian population is now concentrated in cities:[3]

Table 2
Urban Population, Socialist Republic of Macedonia, 1991

City	Population
Skopje (capital)	448, 229
Bitola	84,002
Prilep	70,152
Kumanovo	69,231
Tetovo	51,472
Tito Veles	47,326
Ohrid	42,908
Stip	42,826
Struminca	34,396

MACEDONIA CONNECTS TO THE PAST

Macedonia initially wanted to remain part of the former Yugoslavia. In 1991, Macedonia's President Kiro Gligorov and Bosnian President Alija Izetbegovic, in trying to avoid a disastrous outcome of the disintegration of Yugoslavia, suggested a loosely organised community of former Yugoslavian republics, similar to the EU model. Macedonia fought to remain within such a system and to remain within Yugoslavia. However, because of the war in the former Yugoslavia, the only option at that time would have been

Figure 1.3
Map of Yugoslavia before 1992

to join in a Greater Serbia, a possibility the Macedonians declined. However, once independent, Macedonia regained an image of itself that was celebrated extensively in 1993 through several anniversary events, such as the100-year anniversary of the founding of the Inner Macedonian Revolutionary Organisation (IMRO) and the 90-year anniversary of the Illinden Uprising of August 2, 1903. (August 2 is the religious name day of St. Illija, hence Illinden.) August 2 bears historical significance in Macedonia's history because of another event as well: the August 2, 1944 meeting of the ASNOM (Anti-Fascist Assembly for the National Liberation of Macedonia), at which the fate of post-war Macedonia was decided.

These three anniversaries have fundamental meaning for Macedonia. The Illinden Uprising originated as a short-lived rebellion against Ottoman rule, with the Krushevo Republic being declared for 10 days and adopting a modern manifesto that announced democracy, human rights, religious freedom, and ethnic tolerance. The ASNOM meeting is celebrated in Macedonia for its emphasis on the Macedonian language and national identity, without which today's republic would have been unthinkable. Historically, there was another event of importance for today's Macedonia: the celebration of 1100 years since St. Clement was declared bishop. St. Clement was a famous student of the Slav apostles of Cyril and Methodius. He founded the first Slavic university in Ohrid and, through that institution, began the study of Slavic-Macedonian language and culture. The archbishopric of Ohrid existed until 1767, when Greek clergy in Constantinople insisted that this church was illegitimate. Only in 1958, with the co-operation of Marshall Tito, was the Macedonian Orthodox Church founded again, but it has never been recognised by the Serbian Orthodox Church.

This list of events provides an opportunity to remember that Macedonia's current national sovereignty did not emerge in a void, and, more importantly, that the groundwork for this sovereignty was laid during its Yugoslavian past. Nevertheless, despite these national events, the fall of Yugoslavia left a personal void for many people within the Republic, one that could not be easily filled.

WITHIN: MACEDONIAN IDENTITY

In 1991, there was confusion and fear; people were concerned that war would erupt at any moment. I was even asked to leave by my hosts, who feared that something might happened to me. When New Year's arrived and some people decided to have fireworks, several people ran for shelter convinced that war had started. In 1992, some normality returned, and people started to accept the changes around them and to settle into a routine again. However, there were two main irritants, especially for the young people in Skopje, which slowly replaced the fear of war: 1) they were not free to travel where they wanted to, and 2) Tito had become discredited.

In the summer of 1992, I and my friends, a group of 22-year-old engineering students, took a bus to Ohrid. It was hot, and the bus was winding its way up the rocky mountain roads full of young people going on holiday, not to Greece, as in 1990, but to the small tourist town of Ohrid, which was to become so populated during the summer that the lake became contaminated and a health warning was issued. The windows of the bus were open, a breeze was blowing through, and some young men were playing guitars and singing old Yugoslav pop songs:

> I have dreamt last night, that I don't have you,
> that I lie awake on a snow sheet and quietly
> some other woman is calling my name through the night—a bad dream.
> I have seen in my dreams a white lily, black horses, and wedding guests
> without a song and quietly, without a sound going somewhere, are some
> dear people, but where?
> A bad dream....
>
> Bijelo Dugme, 1976

Spirits were low: friends and colleagues from Sarajevo were not going for a holiday. Life was different this year.

Summer had been the most important time in their lives for as long as my friends could remember. As soon as the summer holidays were over and the young people returned to their studies or jobs, they would begin talking about where they planned to go next year and share stories of great holiday spots and those to be missed. This year, everybody would be in Ohrid, even though the town's inhabitants were not happy about the invasion of *Skopjanici*.[4] There would be no foreign tourists this year. Anja whispered in my ear, "I would like to travel very, very much." When asked if the political circumstances made her afraid that she would not be able, physically or financially, to leave the country, she turned to me and said, "If the standard of living was the same, not travelling wouldn't be an issue. Maybe I would like once or twice in my life to go somewhere and see the world. Not somewhere too far, not too expensive."

My friends in Macedonia define freedom as having possibilities. With the disintegration of Yugoslavia, Macedonians felt that possibilities had been taken from them. Looking at their lives with this loss of freedom in mind, people within Macedonia slowly gained an understanding of what Macedonia was without Yugoslavia. As Anja recalls, in 1992, there was still the very real threat of war:

> There was a time, I don't know if you remember, when we could hear shooting and I thought that I should go into hiding. In the hiding place, there are cockroaches and rats and the darkness is disgusting. You can imagine how it feels to have a cockroach on your skin: it is terrible in the darkness. I don't watch the news at all and when they show all those pictures [of the war in Bosnia] I do not want to watch it. I cry a lot. I cry when I see these things and think what may happen to me. And if I think about all those things, what would it change if they recognise us. Nothing would change even if they [the EU] recognise us I think I could not feel anything, but maybe with time it would change.

In 1992, fear overshadowed any concern about being recognised as a country. For many of the young Macedonians in Skopje, whom I met during this time, the question of whether Macedonia existed was too abstract; it was not real to them. Everyday life involved surviving another day of real fear, anxiety, and bleakness. This fearful, anxious life was still contrasted to life in socialist Yugoslavia. While I sat in Maja's apartment looking at her edition of Tito's biography and drinking coffee one hot summer's day, Maja brought her pre-independence life and her post-independence life together for me in the following reflection:

> My life has always been the same I think: we have always lived in an unstable country. During certain periods of time life is more difficult and then there are periods that are less difficult. Our life is still unstable, especially the economic situation. There were times when the economy was better, but there would be periods of crisis and life would be more difficult. Now I think the only things we find difficult are not being able to travel and having to hate people.

It is important to note that, in this statement, the economic situation is blamed for causing instability. The disintegration of Yugoslavia derived from a dispute over resource allocation between Croatia and Slovenia on one side and the central government in Serbia on the other. The more prosperous republics of Slovenia and Croatia felt that too many of their resources were allocated to the poorer republics of Macedonia and Kosovo and to Serbia. But it was the notion of ethnic hatred that become omnipotent in expressing the conflict in Yugoslavia and, today, inside of Macedonia. Maja and her friends struggled to understand what they saw on television during this time: Sarajevo in flames and unimaginable atrocities conducted by people they held dear. They had great difficulty understanding the abyss that had opened, and it was hatred that could be used to rationalise the war that had erupted between people who, it had previously been believed, were brothers. Today within Macedonia, it is again hatred that is summoned when trying to rationalise the conflict between the country's two main ethnic groups. Interestingly, however, Macedonians experience tension between the hatred that characterises this ethnic conflict and the knowledge that one of the ways to please the EU and gain access to greater economic prosperity is to present Macedonia as liberal and tolerant. Although actions and attitudes within the country, particularly towards the Albanian population were far from tolerant in 1992, every chance was taken to demonstrate the efforts put into avoiding ethnic conflict and how Macedonia was the best example of a liberal, tolerant social democracy. Draga, a middle-aged mother of two and a teacher, understood this need to demonstrate tolerance, and it prompted her answer to my question about whether she fears the same ethnic violence in Macedonia as in Bosnia: "I think we are the only ones in the world that can live with other nationalities and that is how we have survived. We have no outstanding feeling of nationality. That's just our life."

The choice of political system for Macedonia, then, was made through a personification of Macedonia as a fellow European democracy, within the social democratic system of the European Community. As such, Macedonia's socialist sensibility was used to gain access to European democracy, in contrast to other Eastern European countries such as Albania.[5] It was not "fend for yourself," but "fend for the weak." This socialist sensibility became more important throughout the years as it preserved self-worth and an identification with socialist Yugoslavia. Socialism was not banned in people's personal lives; rather, it was seen as a superior social tool that had to be improved in only two ways: first, any totalitarian sentiment had to be eradicated from the political, legal, and economic arena and, second, a modern European market economy had to be introduced to replace the weakest points in the socialist system. In 1994, this ideology was endorsed through statements such as "We have Turkish and Albanian neighbours and all of them are living very well, helping each other." "You know if your stomach and pocket are full, there is no need for hate," a young mother of two told me in 1994. But people were aware that Macedonia was not only receiving European aid in order to join the European community quickly and safely, but that Europe, in turn, formulated this welcome in a relationship of dependence. A woman in her late fifties describes how she saw the relationship between Macedonia and Europe:

> We are used to a poor life. When Europe condemns us to such a life we must not fight them. I still believe in civilised Europe; today or tomorrow they will have to understand that although few people live here, we are in no way inferior to other civilised people in Europe. We are Macedonians: that's first. But we are international, refined, and educated.

In day-to-day life, people constantly reflected on the difference between educated Europe and the uncivilised Balkans. It is this preoccupation, this fear of being seen as inferior to Europe, that is the determining force for independent Macedonia. Within this preoccupation lies the dominant feature of the Macedonians' search for self-identity—they want to decide both who they want to be and who they don't want to be. Answering the question "Do you feel Macedonian?", Ana, a 45-year-old engineer, reveals this fear of being considered inferior:

> What's the point of it? The feeling of nationality doesn't matter, when I don't have anything to live on. I believed in Yugoslavia, but now I don't. What kind of country is it, if it doesn't care for us? We are university-educated, we are civilised and informed. We are not children.

Education is important in Macedonia: it distinguishes European countries from the uncivilised Balkans. It was Tito who brought education to Macedonians, and the old people remind the children of this. It is education's transformative power that gives self-worth and changes an uncivilised Balkan country into a European one. Macedonians are considered well educated, especially by my informants, most of whom have finished university degrees in engineering. During the Yugoslav years, Macedonia was considered part of a wider scientific community, but, with the political changes of 1991, this inclusion ended, and many felt excluded from the rest of the world. Books were not readily available and neither was current knowledge of the surrounding world. As a cosmetician[6] told me, "I am Macedonian, but I feel European. It is very hard for me when they separated us and put us back, but we should be in Europe—our place is there." "Putting us back" was a phrase used mostly by older people to refer to the idea that Macedonia had been elevated from a typical peasant society to the status of a European country because of the Yugoslav years. Many felt that the attitude of the outside world, evident in the refusal of Greece and other countries to recognise Macedonia, was directly linked to attempts to return Macedonia to its status as a backward peasant Balkan country. It did not help that Turkey was one of the first countries to recognise Macedonia, since its acceptance was seen as the country assuming the role of Ottoman master again and reducing Macedonia to its peasant past. A bitter saying, repeatedly quoted even before 1990, was "What do you expect, we were peasants for 500 years under Ottoman rule." The saying suggests that Macedonians consider the period of Ottoman rule part of a dark history that kept Macedonia separate from Europe for a long time, and it was therefore ironic, and somewhat unwelcome, that Turkey was the first nation to recognise Macedonia formally. The fact that Turkey recognised Macedonia deepened negative and chauvinistic feelings toward the Islamic Albanian population, which was seen by the Slav population as holding Macedonia back; the Islamic Albanian minority's backwardness was seen as an obstacle to joining Europe. Today, this sentiment is at the heart of the ethnic conflict that threatens to engulf Macedonia.

Over the years, issues of self-worth, both personal and national, dominated many of my conversations in Macedonia. After the disintegration of Yugoslavia, there seemed to be a struggle to reassess feelings of self-worth:

> I am sorry that Yugoslavia is gone. I personally recognise the old Yugoslavia, but I am also a typical Macedonian in the sense that I was born in Skopje. Macedonians are worth something. I come from this country, and I am not ashamed that we are seen as backward in comparison to some other European countries. I am Tito's child, and he was the man who taught us that we must withstand, we must avoid temptations and go on. We studied, we were children of average parents, but we made an effort and we went ahead. We believe in education: this is the weapon with which we can get out of the poor life. This we learnt in Yugoslavia. We have to continue to progress and learn things even if they are not ours. We have to learn to accept.

In this context, I understand "accept" to mean "being accepted and accepting" and, more specifically, suppressing feelings of hatred for the Albanian population, at least in official discourse. And so, Macedonia presents itself to the outside world as being worthy of a place at the heart of Europe. Svetle, visiting me one morning for a cup of coffee before going off to buy some perfume in a duty-free shop, formulated it in this way: "If Europe accepts us it will be very easy for us to immediately feel we are Europeans because for 50 years we lived as Yugoslavian. We have no inferiority complexes."

The ideals, which are understood to be distinctively Macedonian, are directly related to Tito's achievements. The ideals that Tito planted in Macedonian society, then, seem to have legitimised Macedonia's desire to be an integral part of Europe. Nevertheless, in 1992, Tito became discredited, and the governing party, VMRO, which advocated nationalism as a solution for Macedonia, took action to remove any overt association with Tito. It became one of VMRO's policies to take down Tito's pictures in all public buildings. This action angered many Macedonians:

> I regret that the nationalists took Tito's picture down. I consider him very important for Macedonia, no matter how others neglect him. While he was our president, Macedonia became a legally recognised state for the first time. He was an exceptional person, rare and despite the totalitarianism, a legend of his time.

A 23-year-old engineering student also criticised Macedonia's official renunciation of Tito:

> I liked Tito. Well, I was a child and we waved at him—now they attack him. I don't like this. He did good things; now they present him as having never acted for the public well-being. They associate sordid stories with him to sully his name, fabricating stories that he had 16 children and so on. To do this is not good for us. We waved at him, not only us children, but everybody waved at him, and now they act like we had no association with him.

Her opinion was widely shared amongst other young students, and by their parents. The mother of a friend, who was married to a Yugoslavian diplomat, expresses Tito's significance to Macedonia's identity in this way:

> I do not miss Tito or Yugoslavia. It is over. I do not even think about Tito. But we are very grateful to him, as he changed our life very much. For the first time, we had our own country and our own identity. Before we were nothing, and now they try to do the same again to us, to make us nothing. Somebody said you are Greek, you are Bulgarian, you are Serbian, but now we are Macedonian. And Tito said, "You are Macedonian, you are a nation." If you still see his pictures on the walls, it is because we are grateful for that. If not for him, we would not have our own country. We would be still Bulgarian, Serbian, or Greek.

Many shop owners started putting up pictures of Tito in their shops in 1992, thus combining the new privatisation policy of the market economy with old socialist ideals. The issue of Macedonia's political inheritance and its current and future allegiances became a concern for all, because, as an informant, Biljana, points out historical truths, while difficult to discern, affect lives:

> You know, after all that happens here, I am totally confused. Now I do not know anything about our history, because something that is told about Tito three years ago is said differently now. Maybe it is true, maybe it was true, and maybe now it is true. Today everybody is interested in politics not because of politics, but because of life. It affects everybody's everyday life.

All my informants were aware that Macedonia's fate depended almost exclusively on its political future. For example, during my research President Kiro Gligorov was understood to be a decisive figure—and nearly as much a father figure as Tito had been. Kiro Gligorov—in the seventies President of the Parliament of the Socialist Federal Republic of Yugoslavia, and then integrated into the pre-mortem government of Ante Markovic, to implement a new market economy in Yugoslavia, leading to his election as first president of independent Macedonia (until 1999), mirrors Macedonia's transformation from Yugoslav-Macedonia to European-Macedonia. Suse sums up the general understanding of Macedonia and being Macedonian:

> I feel Macedonian now. I felt Yugoslavian two years ago because somebody told me that I was Yugoslavian since I lived in Yugoslavia. Now I am Macedonian, not only because I live in Macedonia, but because there is no Yugoslavia, and there never can be again, ever.

In order to survive, Macedonia had to be redefined along new lines. The determination of these lines, however, is still under dispute essentially because the redefinition of Europe is also underway. Macedonian identity is directly linked to the imagery of Europe and the Western World, and, although people have quite different memories of the past, they always want Macedonia's future to be in Europe in order to recover what they feel they lost with the disintegration of Yugoslavia. The loss of economic standing and their inability to travel have had a direct impact on the way Macedonians feel about themselves and how they think others view them. Here is Lydia, a 25-year-old engineering student:

> I miss Yugoslavia. We had a better standard of living. Now, wherever I go, people look at me as a potential refugee or someone who will cause trouble, a lunatic Balkan person. But when we were in Yugoslavia, people respected us, and we stuck together. I miss Yugoslavia, I miss the music, I miss the people, and the relationships we had. I was in Austria this summer, and we met people from all over Europe, and the first day we met we discovered that we are now Croatians, Slovenians—former Yugoslavians … we stayed together until the last day. I felt these people were closer to me than, for instance, people from Bulgaria. We love the same sports, the same sport heroes, the same music, the same dances, and we went on holiday in the same places. I have never been to Bulgaria, for example, but many times to Slovenia. I know so many places in Slovenia, and that is what I miss. Now it is all different. They study their own history, their own way of living; they have their own money. We do something else. Now I cannot travel.

A colleague of Lydia's, Sonja, says, "I would like to go abroad, anywhere, to gain some work experience. I have always had the chance to travel; it is in my blood. I would like to continue to travel, to meet other people, learn about other cultures, see their architecture, and learn about their history. That is one of my biggest dreams in my life." Her mother, too, reflects on the changes:

> I cannot comprehend what is going on, and I am really sorry about what is happening. I grew up and lived in Yugoslavia. I studied in Ljubljana. I have friends from Dalmatia, Bosnia, Serbia; there was no difference between us, and we had so many things in common. I felt nothing could separate us, and I can't understand what happened. It is awful, terrible; nothing more devastating could have happened. I am sorry even now that Yugoslavia fell apart. Yugoslavia was small to me, not to mention Macedonia now ...

A Turkish woman from Berevo, who had worn the veil until her husband tore it away from her against her will after attending a communist league meeting, also recalls Yugoslavia nostalgically:

> We older people will never forget Yugoslavia, but the younger will. But we won't forget that we could go freely wherever we wanted to go, and now you have to be scared to cross the borders of even Macedonia. One travels with fear. Now it is not free. Tito was an influence for 40 years, and we lived in freedom. We were proud of Tito, but Yugoslavia is gone now, and it should have been better for the young people. Our life is over, but it should have been better for the young.

The fear that the children will suffer, even though the parents have worked all their lives to give them a better life than they themselves had, runs deep in Macedonian society. An old woman from Prilep expresses this fear:

> This will just impoverish us again. The young people do not know what poverty is, but I grew up before Yugoslavia existed, and I know what poverty is. Now the young people will go where we came from; that is what Macedonia was about. Macedonia is nothing without the old Yugoslavia, but we do not want the Serbians here again either. I am too old. I cannot say anything anymore.

Her young niece comments on the deficiencies of the present as well:

> I am not interested in politics, but before, under socialism, we had more opportunities. We were given the opportunity to learn; it opened our eyes, unlike the changes we are now undergoing. Everyone was employed, had a flat, a car, furniture, money for living, for holidays, clothes; now it is not like that ...

The near impossibility of obtaining a visa and the concomitant inability to travel are especially traumatic experiences: "Not to be allowed to go somewhere and see something, there is nothing more difficult than that," Jasmina, a 25-year-old engineering student from Ohrid, confesses.

CONCLUSION

The personal void many Macedonians felt after independence is directly related to the fact that Macedonia's identity is strongly connected to Yugoslavia, despite a thousand-year history outside of this country. Some nationalist and expatriate Macedonians try to fill this void with nationalist rhetoric and a glamorous ancestor. Others expressed their beliefs in what is, ultimately, a desire to be European. As one informant said, "If we all get together again, then I will be Yugoslavian again. Now I am Macedonian because this is my country." "I think what makes someone feel Macedonian is when you feel you are part of it," another stated. "I felt part of Yugoslavia, and I was Yugoslavian. When I am part of Europe, I will be European." Essentially, their quest for identity is a quest to be European.

Similarly to Brown (1995) and Danforth (1995), Poulton argues that nationalism determines Macedonians' very being:

> The impact of modern nationalism on Macedonia and its peoples has been momentous. It is one of the prime reasons for the area becoming the "apple of discord" in the Balkans and the centre of such intense controversy.... Nationalism ... is seen as an activist ideological movement which aims to unite all members of a given people on the basis of a putative shared culture. As such it claims to represent the whole collective, however defined, and is antagonistic to competing cultural claims on the totality or parts of this collective, which is deemed by the adherents to constitute an actual or potential nation. (Poulton 1995: 6)

The everyday life I observed in Macedonia between 1991 to 1996 did not support such a statement, but today's ethnic conflict might just confirm it. However, after the disintegration of Yugoslavia, people in Macedonia concentrated on their individual lives rather than on political events, which does not mean that people in Macedonia are apolitical. On the contrary, people in Macedonia have appropriated politics into their everyday life and given it their own distinctive meaning, leaning, as they do, on the hope that one day they will be full members of Europe, accepted, not as a nation, but as a people. But it is Europe that today determines the course of Macedonia's fate. And, as it did in Kosovo, Europe might just fail to live up to its promise. There are many moderate voices in Macedonia, Albanian and Macedonian alike, but they are less and less heard. My research has shown me a country with great potential—with ethnic conflict, but also with the willingness to overcome distrust and differences.

To understand how the idea of being European creates ethnic tension in Macedonia, I would like to point the reader to the fact that cultural change is generated by encounters between local and global conditions. We can see this when we look at events that occurred during the disintegration of Yugoslavia and the subsequent social transformation of the now independent Republic of Macedonia. People within Macedonia may have reacted variously to external challenges to Macedonia's political identity, but everyone did react. For example, there were the animosities within Macedonia between the Albanian and Macedonian population, which have resulted in armed conflict, to the surprise, I believe, of both parties. My Slavic informants explained these animosities quickly: the Albanian population was seen as a reminder of the historical inheritance of 500 years of Ottoman rule. This inheritance was defined as conferring a strong patriarchal and agricultural past. This image reflected what urban Macedonians, rightfully in their minds, aspired to transcend, in order to be on equal footing with Western Europe. This claim of equality was justified through reference to Macedonia's unusually high level of urbanisation, which has been credited to Tito and his efforts to modernise Macedonia and to make Macedonia a proud partner in his creation of the Yugoslavian state.

Today, the republics of the former Yugoslavia show very different political, economic, and social conditions, just fifteen years after the armed secessionist struggle began in 1991. The people in Macedonia today have realised that they have to stand on their own. This realisation came only after a long and painful process, and, because of this process, Macedonia differs from most of the other republics that have, more or less consciously, chosen independence. Recent fighting in Macedonia, where the Macedonian military jumped at the chance to prove that there was militancy in the Albanian population (which has fallen victim to a small group of extremist KLA fighters from Kosovo), demonstrates the plight of the Macedonian people: they have become increasingly polarised and are in economic straits, and, consequently, seem to be on a difficult road when it comes to becoming recognised as a viable state in Europe.

Ethnic conflict in Macedonia, then, is possible not because of ethnic hatred, economic distress, or religious intolerance, but because of the division created between what is termed "Balkan" and what is termed "West," because of something Maria Todorova has aptly defined as the Balkanisation of the Balkans—a variant of Orientalism (Todorova 1993).

In short, we are presented with very specific images regarding Western and Eastern Europe, dividing the world into economically developed and under-developed parts of the world, a world that, in the cold war, had been held apart very usefully by the two idioms of capitalism and socialism. These images give us a familiar matrix of exclusion, of "Otherness" operating around the "Us and Them" paradigm that is still in effect today. And it is this "Otherness" that has been adapted by many of my Macedonian friends. In an effort to make sense of their exclusion, they have translated "Otherness" into their own world, where the Albanian population becomes the people seen as "Others" by Europe, and from which my friends differ, proving this through their personal interactions and life choices.

This self-reflection, this examination of one's own normative position, demonstrates that Western scholars have overlooked a very vital understanding of what happened when Yugoslavia fell apart—the appropriation of the images of "West" and "East" as they are used by the Western media, an appropriation that introduced a fundamental change in the perception of "Otherness" within former Yugoslavia, in turn altering historical memory.

Memory is made up of the simultaneous processes of forgetting and remembering that have been particularly evident in the conflict around Macedonia, when nationalists (first, to a great extent the Diaspora) ventured back to Alexander the Great and Christian Orthodoxy to find Macedonian identity and, by doing so, excluded a great part of its nationals, namely, the Albanian population within Macedonia. This nationalist effort to create a Macedonian identity and a Macedonian past was hampered, though, by the fact that Macedonia has to deal with two different pasts, which needed to be separated, as shown in Chapter Two, the rural past and the Yugoslavian past. Both of these pasts have very different meanings. Today in Macedonia, we can see two currents when Macedonians consider these two different pasts. When they consider the rural past, people tend to forget the unity that existed in Yugoslavia and concentrate on nationalism. Western media and academics often argue in response to this that religion, which was suppressed under socialism, resurfaced and explains current ethnic conflict.

I, however, argue that, in fact, very specific images of "West" and "Balkan" inside of Macedonia cause a conflict because of who is seen as occupying the rural past. In other words, the Albanian population is seen as the "Balkan," occupying the "rural past." When we, in turn, consider the Yugoslavian past, we find people remembering the unity they once lived in. It is an interesting fact that, when Kiro Gligorov survived an assassination attempt in 1995, rumours circulated that the Bulgarian Mafia was behind this attack, but to my surprise given the already tense relationship between Macedonians and Albanians, Albanians were never mentioned as possible culprits. The enemy, then, was sought outside the borders of Macedonia, and the unity of Macedonians and Albanians during Yugoslavia was remembered instead.

If this unity is remembered, why then are we looking at the possibility of another armed conflict in a former Yugoslavian Republic? Here, economics and the difference between rich and poor are used by the West to explain ethnic conflict. However, the conflict arises because Macedonians see Europe equating their country with the "Balkans" rather than the "West."

When considering Macedonia's two pasts, we must realise that both of these "pasts" create very specific notions of sameness and difference. We have to make a distinction between Yugoslav nostalgia (unity) and ethnic hatred. There has always been conflict between the Macedonian and Albanian population in Macedonia. However, the meaning of this conflict changed through the Balkanisation of the Balkans by the West.

How did the memory of difference and sameness change within Macedonia? Let me give you an example from 1988. On a beautiful Sunday afternoon I go for a walk up Vodno, the mountains towering over Skopje, with a friend of mine. We have a beautiful, leisurely walk, not fearing god and the world, listening to the sound of crickets and talking about childhood memories: playing cowboy and Indians, or, in Stojan's case, partisans and Germans. Suddenly we hear a slight noise of bells, and the posture of my gentle friend changes immediately. He starts swearing that he has forgotten his knife and that we are in great, great trouble. In 1988, I was still wondering what one could possibly fear in the mountains of Yugoslavia. My friend quickly explains that we will probably encounter Albanians. Not understanding why this is worrisome, I concentrate on how my friend could possibly expect a horde of Albanians in the midst of the mountains and on why this horde would be so kind as to ring bells before an attack. Stojan, meanwhile, makes a great effort of unearthing a great big stick that would have to do for our defence. Then we turn a corner to see an old, tired goat chewing some twigs standing in our way guarded by an even older, very crooked, wrinkly man, who could be identified as Albanian by the cap on his head and, probably, though I do not remember, by his white socks. All this might sound quite funny, and, as Stojan's friend, I did chuckle at his sudden attempt to be the protector of the tender maid, which fitted with our childhood remembrances. Nevertheless, in deeper thought, I recognised, and would encounter more and more often, the deep-seated distrust between Slavic Macedonians and Albanians. (As the old man was probably close to a heart attack upon having this young Macedonian man appear around the corner.)

That same evening, we went to the cafés of Stara Carsija, in the old part of town, which was crowded with young Macedonians. Though I cannot claim to have seen any Albanians mixing in this crowd, it is important to note that Stara Carsija is in an area easily identified as the Albanian part of town during the day. Probably, somewhere in between the fashionable cafés of the young Macedonians, the old man from the mountain was recounting his adventure to some of his Albanian friends in one of the many "Turkish" coffee shops. The point I want to make is that, today, Stara Carsija is empty at night. No young Macedonian, except some who belong to the alternative scene, can be seen between the old houses. I insist that this is no coincidence but an indication of a change in the relationships between Macedonians and Albanians during the last ten years. Fluid borders are played down, and new borders are created, such as the demand for a solely Albanian-speaking university in Tetovo. This changed animosity, this emphasis on absolute difference, is called for by Western Europe's insistence on what "Europe" is. One of the foremost definitions is that Europe is not Muslim. (See the discussion on Turkey's acceptance to the EU.) Consequently, religion becomes newly emphasised.

These complexities show that we cannot actually talk about forgetting the unity of Yugoslavia or remembering a rural past, but instead should talk about memory and forgetting as oscillating, meaning that, sometimes, memory is omitted. This omission, in turn, is very much influenced by how the Balkan is imagined by the West.

Consider in support of this point two events that happened simultaneously during 1992, each reflecting opposite expressions of identity. First, on August 2, 1992 in Krushevo, there was the re-enactment of the 1903 battle between Macedonians and the Ottoman Administration, which gave Macedonia ten days of independence. In contrast, if you visited Skopje in 1992, you would find more public and private displays of Tito's picture than in 1988, a circumstance suggesting nostalgia for Macedonian unity achieved within the former Yugoslavia.

On what, then, is the discourse of sameness and difference in Macedonia based? When citizens of Macedonia consider the fate of Bosnia and Serbia, they see the Yugoslav years as representing a more glorious past, but this past emphasises unity and brotherhood between Slavic Macedonians and Albanians. Therefore, my informants see the Albanian population as holding Macedonia back from Europe although unity and brotherhood with this population is a "European trait" endorsed by the both their government and the EU. Yugoslav nostalgia, especially in considering the fate of Bosnia and Serbia, is seen as a more glorious past, but this is also seen as a past that emphasises unity and brotherhood between Macedonians and Albanians. As such the Albanian population is seen to hold Macedonia back from Europe, but unity and brotherhood with the Albanian population is also seen as a "European trait" and is endorsed.

The negative image of the Albanian population is caused by a specific imagery within the Western world: The Western world is an urban world in which having an education is seen as being cultured, "*kulturna.*" In this world, consumerism and freedom, according to Western ideology, guarantees progress. This progress, however, is seen as being held back by the Albanian population and by Macedonia's pre-socialist peasant "*seljak*"-past. However, the state needs unity, and there is a real fear of civil war.

Today, Macedonia fears rejection by the "West" as "Balkan." I believe this explains the violent riots against an Albanian influx as demonstrated during the Kosovo crisis when Albanian refugees from Kosovo were secretly loaded into buses in the middle of the night and transported from Macedonian refugee camps to Albania, an act which tore families apart and met heavy opposition from Europe. Riots in Skopje against NATO air strikes in Serbia, on the other hand, showed a certain Yugoslav nostalgia. However, the nationalists' rhetoric of being a guest in their own country (of becoming outnumbered by Albanians because of their higher birth rate) and their fear of being equated with or of becoming Albanian (and thus "Balkan") leads to a fear that Macedonia might not have a European future. This question of identity, I believe, will essentially decide whether there will be a civil war. Will Macedonians resolve these ambiguities or endorse them? Will Macedonia have a future? My ethnography, which investigates identity, deals to a great extent with what people felt they had lost and what people did to compensate for the loss—a loss resulting from the disintegration of Yugoslavia. I argue that my informants in Macedonia felt a great sense of loss of agency (for example, loss of the ability to travel and to work in a position of their choosing). And in order to gain a new agency, they assimilated the images of "West" and "Balkan" as introduced by the West (in its dealing with the ongoing war in Bosnia). Confidently, they positioned themselves within "the West" and put aside "Albanians" and "Balkan mentalities" (such as violence and corruption) by associating these with images of the "Balkans." Interestingly, at the same time, on the national level, what was Balkan became connected to a rural Macedonian past that the nationalist tried to re-claim, whereas the "West" became connected to the Yugoslavian past, especially by the younger generation.

In this way, Yugoslav nostalgia derives from the fact that Yugoslavia is perceived to have been in closer proximity to Europe than Macedonia is today, especially by the younger generation, who were accustomed to travelling freely within Europe when part of Yugoslavia. Also, Yugoslavia presented a unity that was seen as the antithesis of war. One fact is really important to understand in this argument. The boundaries that are seen as hindering free access to Europe (through visa regulations, for example) are perceived as having been erected by the West in order to keep the Balkans *out* of Europe. Consequently, many Macedonians interpret the Yugoslav past as the ideal future, a future with open borders, unity, political recognition, economic and scientific exchange, and social democracy (not the demise of socialism!). The problem facing this future is that "The West" is attempting to create Macedonia as "non-European." The conflict today in Macedonia seems to prove the country's non-European nature once again. However, I argue that this conflict cannot be termed "ethnic nationalism" or anything similar. Such "ethnic nationalism" is not possible because of the conflicting images of the Balkans and the West, because going back to the past, back to Macedonia's roots, is a move that would create "sameness" with the Albanian population, a sameness that would be seen as equating Macedonia with the Balkans. However, the Macedonians' reaction to this, because of Europe's closed borders, is that they are more inclined than ever to eradicate what is seen as holding them back. The Macedonian Army's salvos on Albanian houses and cars should not be seen as innocent mistakes. In a kind of twist, the KLA rebels who claim to fight for the Macedonian Albanians harm those they purport to help, giving the Macedonian nationalist government reason to expel all Macedonian-Albanians as "Albanian terrorists." European leaders are travelling to Skopje, trying to bring Macedonians and Albanians to the negotiating table, but they overlook some fundamental truths about the situation in Macedonia: although there is clearly heavy discrimination against the Albanian population, this discrimination is not endorsed legally, so the arguments of the KLA do not serve anyone, but only further the organisation's own perverse power games. It is also a fact that, politically, the Albanians, if united, would have had enough representation to change any inherent legal, social, economic, and political discrimination against the Albanian population. A violent conflict is not necessary, and the only people whom this conflict serves are those radical Macedonians wishing to send all the Albanians back where they belong—to Albania, in their minds, though many Albanian families have lived in Macedonia just as long as the Slavic Macedonians. To emphasise, the conflict that presents itself in Macedonia today cannot be termed an ethnic conflict derived from ancient hatred. The Balkanisation of the Balkans by the West creates this conflict. Therefore, looking at Yugoslavia's disintegration necessitates looking at the interests of Europe. We cannot simply argue that Yugoslavia fell apart because of ethnic nationalism, ancient hatreds, its internal structure, constitutional tinkering, internal economic differences, and the end of the cold war. We need an integrated analysis here, one that considers a far more global perspective.

NOTES

1.

Political Party	November 1993	June 1994
Social Democratic Union	22.8%	18.8%
Party of Democratic Prosperity	18.4%	15.9%
Democratic Party	6.3%	8.3%
VMRO/DPMNE National Party	11.2%	8.2%
Liberal Party	3.2%	3.9%
Socialist Party	3.3%	2.8%
Labour Party	2.9%	1.6%

Source: *Brima Gallup Report*, 1994.

2. Office of Statistics, 1991.
3. Office of Statistics, 1991.
4. *Skopjanici* means citizens from Skopje, the capital city of Macedonia. See Chapter 10.
5. See Lemel 2000.
6. A highly educated businesswoman should not be confused with the Western stereotype of a cosmetician.

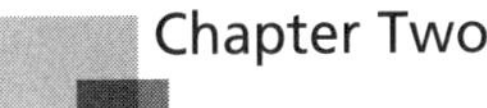

Chapter Two

Mapping Urban Identity

Map of Macedonia

Would it be possible with such aloofness,
with two or three talkative lines and
with two or three conversational colours,
to splash on a canvas the idea of
one's native land?

Well yes, it would be possible!

Our destiny is framed
like something turned upside down
doubled up in pain the horizons of hope,
crushed the visions labelled refuge,
Truth spread out on the palm of the hand
and the pictorial perfection of no way out.

But all you see is a compass and a map
and many pencils scattered about it,
sharpened if not smoothed,
so they can prove themselves in a proud role:
looking at lines we look at grief,
we look at our history—

What we have framed here is not just a sigh
preserved in another colour—

That, too, would be possible; quite possible.

Gane Todorovski

THE DISCOURSE OF IDENTITY IN THE REPUBLIC OF MACEDONIA: THE EFFECTS OF THE DISINTEGRATION OF YUGOSLAVIA

As the disintegration of socialist Yugoslavia took place, the overarching "Yugoslavian" identity, which had encompassed the diversity within the state, ceased to exist for most people. People were left initially to their own devices in identifying themselves as non-Yugoslavian. Over time, a move from this negative identity of non-Yugoslavian began, and a new identity emerged in the tension-filled vacuum left by the collapse of Yugoslavia.

This new identity is derived from Orthodox Christianity and a land-owning peasant past. However, even though the old socialist state has ceased to exist, the Republic of Macedonia was, during my research, still governed by a socialist party, and, in the urban environment of Skopje, the new rural and Christian national identity is not easily accepted. Furthermore, certain characteristics of the past, such as strong local patriotism, appear to be counter to a multi-ethnic national ideal. A Macedonian identity derived from the pre-socialist past, then, necessarily conflicts with the socialist identity that still is expressed through the adoration of Marshall Tito.

As outlined in Chapter One, Yugoslavia ceased to exist for Macedonia when the Yugoslavian National Army (YNA) withdrew from Macedonia in March 1992 and fears escalated that the situation would develop into a second Bosnia. President Kiro Gligorov and the Bosnian leader Alija Izetbegovic tried to preserve the old state in a federation, but this federation ended on September 8, 1991, with a referendum on the independence of the Republic of Macedonia, as it was then called. The tragic events in Bosnia-Herzegovina led to a drastic change in attitude towards the former Yugoslavia.

In the summer of 1992, my interview question "Who and what are you?" tended to elicit responses such as this: "We were Yugoslavian, but we always have been Macedonians as well. Now we have ceased to be Yugoslavians." By 1995, all my informants called themselves Macedonians. More important, they defined themselves as Orthodox.[1] When I asked why they had declared themselves Orthodox Christians in the 1994 census, I received this typical answer: "Otherwise they [the Albanians] will outnumber us [of Slav origin]." These responses indicate the strong sentiments of Macedonian internal social discourse. The Albanian population was seen as representing the Muslim world, which uses its fertility rate to overpower and eventually eradicate Christian-Orthodox belief. This argument can be traced back to 1389, at least, to the events of the battle of Kosovo Polje in which Serbia was defeated while trying to stop further Ottoman conquest. From this point and into modern times, Macedonia has found itself under Ottoman rule. At present, the Albanian population, because of its Muslim beliefs, is often equated with the Ottoman occupation itself.[2]

Two interesting issues concerning individual identity emerge from my fieldwork: 1) the almost instantaneous identification with Christian Orthodoxy claimed by firm atheists and 2) the denial of an urban identity.

People from elsewhere in Macedonia view the *Skopjanici*[3] as having a distinct identity although this view is not shared by the *Skopjanici* themselves. They define themselves not only according to neighbourhood, profession, and school but also, and even more exclusively, according to the town of their family's origin: Stip, Strumica, Veles, Prilep, or Galichnik. The places they name are their grandparents' and parents' birthplaces even though nearly all of them are first-generation, native-born *Skopjanici.*

It is also apparent that one's ascribed origin naturally confers attributes: women from Prilep are very mean, women from Galichnik are good to marry, people from Skopska Crna Gora are especially prone to black magic, and people from Kriva Palanka are aristocrats.[4] All of these places thus carry a distinct local identity. What these young *Skopjanici* are stating is their inherited connection to a specific rural area of Macedonia with a specific past. It is this rural past that is claimed to be the root of the modern Macedonian state in public discourse. Brown suggests that, in Macedonia's past, an extremely local patriotism rooted in the symbolic spaces of a village and its history without wider extension is the most significant force of identification.

Politically, people also tend to identify themselves with a specific place rather than with a political unit. Many Macedonians feel at odds with their current state and its wider political arena, as they see their politicians working for personal power and material gain. Where the Yugoslavian state has ceased to exist, a personal identity is offered by a specific locality that is given by one's rural past. Furthermore, today the state can only offer identification as a citizen but not as a "people." Only a "nation" or *narod*[5] could represent "people" or "ethnicity." Here Christianity, which dominates most local rural pasts, brings Slav people into a single group, into their *narod*. In this respect, belonging to a rural area of Macedonia is belonging to a particular past that mitigates against isolation as it allows belonging to an extended family, tied by kinship with others to a specific place, a specific land, and united with other people from the land by religion. However, in opposition to this rural identity stands the highly impersonal space of Skopje, which was transformed into a modern city following the earthquake of 1963.

YOUTH IN THE URBAN ENVIRONMENT OF SKOPJE

Skopje, the only urban centre in the Republic of Macedonia, boasts almost half the population of the entire state. When young *Skopjanici* see their origins in the countryside of Macedonia, it is a spiritual belonging to the soil that they present as their identification. The "socialist" world of their parents has ceased to exist with the collapse of Yugoslavia, so the grandparents' generation provides some sense of identity and moral ideals to the young people. Macedonian TV, in documentaries about the past lives of elderly people, presents the idea of *narod*, with ever more frequency over the years. Television informs "its" people of their origins by focusing on the countryside in narrations on its people, folklore, and history told through first-hand accounts of people who lived through this history, thereby invoking the rural identification that is evoked by the public and political discourse.

Most of the social discourse today within the city of Skopje is centred, not surprisingly, on the future. After the disintegration of Yugoslavia, the possibility of a positive future seemed bleak; a living example lay in the events of Bosnia-Herzegovina and Sarajevo.

The film, *Before the Rain* by Milcho Manchevski, nominated for an Academy Award in 1995, gained tremendous popularity in Skopje. In this film, the journalist Aleksander returns from London to his village in the countryside of Macedonia. Skopje is presented only in a series of fleeting images captured whilst Aleksander travels through the city by bus. He eventually comes face to face with the vastness of the countryside, and here the decisive struggle between Albanian and "Macedonian" is fought, with the church presented as being unable to prevent the slaughter. The past is regarded nostalgically as a time when these opposing nationalities co-operated and were even friends: they worked the same land and attended the same schools. Yet, the conclusion of the film is rather depressing: the individual is unable to prevent the cycle of violence, which, as the movie suggests, is not yet completed. At the end, Aleksander is killed by his own people for standing by his Albanian friends. However, he utters the film's important words: "You have to take sides." And his is the character with whom young people empathise. The rain, a metaphor for violence, will come, but the present is a period "before the rain." Today my friends recall that, in the past, there was unity through socialist ideology, which was able to overcome animosities between different people. Young men and women remember waiting for Tito to drive up and wave at all of them. When such an image is evoked, no one talks about "us" and "them." Tito united everyone, and, in Skopje, there are pictures of him everywhere—in Stara Carsija, the old Albanian part of town, and in Stopanska Banka, the modern Macedonian commercial centre. "Under Tito all this would not have happened," is the pervasive sentiment.

Similarly, people recall that, under Tito, the city promised a better life economically than the countryside. It was pointed out to me that many rural villages had no electricity, running water, or flush toilets while city dwellers, because of their better economic situation, could experience the world abroad and take summer vacations outside the country. Many young people have mixed with the "Inter-rail Community," mainly back-packers of Western Europe. Now, for the first time in recent history, access to the Western world has been restricted by visa regulations. The Macedonians feel that these visa limitations label them as second-class citizens in Europe. Unrestricted travel had been unique to Yugoslavs in the "socialist bloc," and the young people took much pride in this fact, as they told me in 1993: "We could travel wherever we wanted. Yugoslavia was very different from communist countries like Bulgaria or Russia. We were socialist.... If the war had not happened, we would have joined the EU very soon." This nostalgia embraces the idea of unity, even though it contradicts Yugoslav

Figure 2.1
The Young and the Old Macedonia

reality and causes more ambivalence in the construction of a specific Macedonian identity. Today more than ever, economic equality with the West is the goal:

> We are going for the money now. Do you know that this year [1994] there are more first year students in the economics faculty than in the engineering? Everybody wanted to study engineering, and now they all want to study economics and become rich very fast. You can only get the girls with a "fancy" car now.

In this world, consumerism has become a major concern, as illustrated in Chapter Six. Computers, Benetton, Levis jeans, and other designer clothes are a must. Prices in Skopje are high and far exceed what is easily affordable on an average income. Nevertheless, people buy these items and consequently endure many financial sacrifices to aspire outwardly to the world of "Beverly Hills." However, their sophisticated literature, art, poetry, and music are largely drawn from previous themes and are not yet imitations of the "Western model." Consequently, there is not an overall denial of the rural by the urban youth and intelligentsia, and the state endorses this preservation and revival of tradition and the propagation of traditional, rural themes.

In this respect, the young people in Skopje are faced with ambiguity in their understanding of who they are and what Macedonia can be. Even though their opposition to the Albanian population defines them as a distinct group of Slavic-Orthodox people, the example of Bosnia and nostalgia for Yugoslav leaves their minds open for co-operating with the Albanian population. They wish to be modern and equal to the West. Not being from the Balkans lets them endorse a mental separation from the Albanian population, which represents for them the "Balkan." However, such a strategy is not an easy one either, as they and the state define their roots in a peasant past that can easily be seen as mirrored by that of the Albanian population.

PEASANTRY

The urban identity of my informants opposes a rural identity. However, there were some very picturesque images of the peasantry found throughout the former Yugoslavia. These positive images, I believe, looked to the past to create a solidarity within the diversity of the Socialist Republic of Yugoslavia: people shared a common rural past that allowed them to feel superior to their bourgeois neighbours, as one of my informants described it. At the same time, a negative image of rural identity originated from looking proudly back from the vantage point of what had been achieved: the victory of modern society. This victory challenged those identifying with peasants, who had simply accepted a fate-driven existence. Young people inevitably had an ambivalent attitude to their peasant origins: on the one hand, their peasant past indicated a backwardness that conflicted with their aspiration to be mobile, modern Europeans. On the other hand, the rural identity was the only personal and social identity that could overcome the alienating effect of the urban environment[6] in which they were living.

They were taught to look back and be proud of the rural past. My informants now picture the life of a peasant as a life lived on small, independent farms worked by members of the extended family. Peasants, they think, once produced everything they needed by themselves and were modest in their needs for comfort. Their wealth consisted of plenty of food[7] and reserves of gold coins,[8] and they had a rich treasure of folk songs and an astonishing number of variations of the *oro*, a round dance that still plays a major role in weddings, birthdays, New Year's celebrations, and party-organised meetings. I was told that those who dance the *oro* and eat plenty of peppers and tomatoes are truly Macedonian and are connected to the community. This view of the peasantry represents a rather cosy picture, one that has to be conveyed through the official "folklore" level of communist discourse.

Even though today's urban youth state their place of origin as the countryside, they still associated negative images with the countryside and the rural lifestyle. For my young informants, these associations derive from their wish for a different, "Western" future and are not related to their own histories or family ties; the negative images are projected definitions of "the Balkan."

THE LANDSCAPE OF THE CITY OF SKOPJE

After the earthquake of 1963, the United Nations along with "Western" states rushed to Tito's aid to restore Skopje. Many of the old buildings still standing were pulled down and replaced with concrete structures. Interestingly, what remained were the Turkish buildings of Kale and Stara Carsija, the mosques, the *Hans* (Turkish "hotels"), the Turkish baths, and Bit Bazaar, the Turkish marketplace with its minarets. Today these buildings are the identity signifier for the Muslim and mostly Albanian population, whereas old "Slav Christian" buildings disappeared, and socialist architecture left nothing specific and reinforcing of identity for the people who came to Skopje after 1963 in the course of rural/urban migration.[9] People were starting to live in the homogenous, identical, and modern apartments that belonged to the socialist state. These apartments represented the state's values, both physically and symbolically, and denied occupants any links within the urban landscape to the villages and small towns from which they had come.

In time, peasants became the *rabotnicki narod*, the working people of the Yugoslavian socialist society, and socialism came to stand for the modern. Modernity meant, in this case, standardised dwellings and stood in opposition to what was believed to have been the past. The negative image

Figure 2.2
Traditional (Albanian) and Modern Skopje

of the rural past and the implications of what it meant to be a peasant were born in the construction of the modern socialist world of Skopje. Skopje was supposed to embody everything that the peasant lifestyle was not.

Skopje street scenes before the earthquake appear to have been similar to those that exist today in many towns like Ohrid, Struga, Bitola, or Strumica. Social life happened in the streets: gossiping and discussions, children playing, men arguing about politics or just sitting in the shade of large trees. Remnants of this life continue to exist in the cafés of Skopje, which hold high social value for the young people, or amidst the table and chess games of men in the park in front of the Trgovski Centre, the older shopping mall, which exists, against the odds of modern urban planning, in the centre of town. The earthquake, though, made it possible to create Skopje as an "ideal" human community, and allowed for the reconstruction of the city as a symbol of the triumph of socialism, which could withstand the "chaotic forces of terrestrial and infernal nature,"[10] such as earthquakes or "unordered" street-life.[11] Landscape became a system of signification and expressed the authority of the socialist state. Intrinsically, the city as landscape became the origin of individual and collective action, the origin of knowing, objectively and subjectively, and the origin of idealist and materialist explanation of the world around.[12] As such, the landscape of Skopje as a city became imposed on its inhabitants and signified the social transition of the country.[13] This internalisation of Skopje's urban space prevailed mainly for first generation *Skopjanici*, those born in Skopje who had never lived a rural life.

When a stranger asks for directions, it becomes apparent how difficult it is to make one's way around the maze of the city. The simple task of supplying oneself with the bare necessities requires organised planning, although different categories of people create their respective maps in the maze of the city.[14] People walk only in very specific areas of the town, and, therefore, the individual does not create his or her own relationship with the city since this relationship is ascribed by group identity. For example, there is not a large intersection of space amongst the Slav, Albanian, and Roma populations of the city.

The city also lacks personal space that could be usurped for social communication and interaction: instead, one can only go "along the streets." A vast area of unused land functions as a refuse dump of sorts. It gave rise to an interesting dispute between, on the one hand, members of different NGOs and, on the other, some of the inhabitants. In the end, the NGO representatives abandoned their attempts to cultivate, clean, and beautify the unused land. Cows continued to graze and trample everything down. Garbage appeared there faster than it could be cleared away. The *Skopjanici* laughed at this desperate beautification attempt, which was eventually subsumed under the rule of the city: people only care for their immediate selves and property since they lack any connection to the space and people around them. As in much post-war building in war-damaged Europe, the city planners chose the cheapest solutions, used concrete very freely, and left little or no green space that residents might have used. Not only the provision of green space but also the city's climate and any accommodation of nature were neglected: people often mention that, in certain areas of the rebuilt city, there is no circulation of wind. As a result, Skopje is turned into a hell of blazing sun and scorching pavements in the summer and is close to uninhabitable:

> I cannot imagine how it will be to work this summer and not be able to leave the city. It is too hot. How am I suppose to work? I have to run around when the day is the hottest, and at night when it cools down, I cannot go out because I have to go early to bed. (Informant 1995)

The only goal of the *Skopjanici* in the summer is to leave the city for the countryside or the coast. In the winter, the fog is recorded as worse each year and renders the city virtually inaccessible by air.

Thus, life in Skopje is a series of isolated events, nodes of activity between home, shopping, recreation, and work, made accessible by a no-man's land of streets. One feature becomes more and more prominent in Skopje's landscape: the shopping mall. It developed from the need to provide a commercial and social focus that people no longer found in the streets. Thus, the city as construct imposes conformity and denies social and environmental diversity.

Adjacent to this world of "modern" Skopje stand the quarters of the Albanian and Roma populations; the division between these two worlds is not only linguistic but also physical. The boundary in the city is drawn by the Vardar River; to the south is "modern" Skopje with the shabby apartment buildings of the Slav-Macedonians, and to the north is the old city—Stara (Old) Carsija, the area of Bit Bazaar, the quarter of the Albanians, and the shantytowns of the Roma. A state document "advertising" the Republic of Macedonia could not avoid commenting on the actual appearance of the "modern" housing: "In the field of housing, however, non-economic relations and all kinds of experiments were predominant for a long time. As a result, there was relatively swift degradation in the newly built housing facilities, while insufficient attention was paid to the fact that depreciation should be countered by renovation."[15] Thus, as degradation occurred, the southern Skopje "lost its soul" while the shanty towns of the Roma and Stara Carsija of the Albanians, even though they run counter to the image of "modern" socialist society, are the liveliest areas of Skopje. These areas have busy street corners and are the centre of a lively black-market economy and money exchange activities. Stara Carsija houses the largest "green market," so named because it is supposed to sell only vegetables and fruit. This market, catering to diverse customers and offering everything from toilet paper to sweaters and jeans from Turkey, showerheads, tools, watches, glue, cigarettes, and alcohol, lies across from the newly built shopping mall. The new shopping mall is frequented mostly by the "modern" *Skopjanici* and offers predominantly consumer goods. While at Bit Bazaar the streets are humming, busy with people, the long floors of the shopping mall are nearly empty and feel sterile in comparison. Interestingly, the government is trying to limit the role of the "green markets." A state enactment in July 1995 confines their sales to only "green" produce, although previously virtually everything could be found at the market at a reasonable price. This enactment was to protect the shopping malls with their private enterprises offering expensive consumer goods. However, there have been demonstrations against this law, claiming that it is discriminatory because these markets are predominantly Albanian enterprises.

The city and its environment establish many of the preconditions for the perception of identity prevalent amongst young people. The limits and challenges the city sets on the mind are embodied in the search for a new definition of oneself and "the other." The young people of Skopje struggle for a particular form of identity that they see emanating from the outside world, much of it derived from Western Europe and American TV and much of it perceived as missing from their city.

The young people tend to identify themselves as "modern" Slavs by their place in relation to the north-south division of the city, with the Slavs living in the south. The placement of a person in the geographical landscape of the city is important in defining an Albanian or Macedonian. To be Albanian suggests, to Slavs, having many children, wearing the veil (if female), purchasing women, participating in criminal activities such as drug cartels, adhering to Islamic fundamentalism, being a foreign worker abroad (a *Gastarbeiter*), and being from the north of the city. In contrast, to be Macedonian is seen as conferring true Yugoslav citizenship, as having the ability to be a tourist abroad, as raising a limited number of children, and as being a modern, socialist person from the south of the city.[16] Being socialist, in this context, means to have risen from the rural past, marked by the "cattle-like behaviour" of the peasant under Ottoman Rule, and to be modern. Yet the city's construction of "ideal modern-socialist" gives its inhabitants a difficult and contested relationship with such terms. Moreover, with the "other" within, the mind is set for an ambiguous definition of the terms "rural past" and "Orthodox Christian" upon which Macedonian identity is built. The terms, like the terms modern and socialist, gain a more twofold character.

THE PAST

Given the urban environment of Skopje and its influence on the perception of inhabitants' identities, examining how the past is reflected in present-day actions and beliefs is vital. Macedonia as a geographical area has

Figure 2.3
Macedonian Village Life

undergone profound demographic changes, which have caused instability and political change, which, in turn, have resulted in shifting allegiances and group identities. In addition, the mountainous area made communication between people difficult and led to a compartmentalisation rather than to a unification of Macedonians. For administrative purposes, the Ottomans created the *millet* system, a system that divided its subjects not according to ethnicity, language, or nationality but by religion. Thus, religion has been one of the main factors in differentiating between various groups. Poulton (1995: 45) writes that, essentially, the Ottoman Empire "was non-assimilative and allowed the separate Balkan peoples to retain their individual cultures and identities—this aided by the geography of the region." The Macedonian socialist state tried to pull these diverse people together by creating a modern city, the city of the *rabotnicki narod*, the working people. However, when the state created this monument to modern socialist society—Skopje—the city lost its rural roots and a sense of belonging. For the people of Skopje, the sense of belonging was evoked only through folklore. The modern city was meant to give its citizens a new identity and a future, whereas the countryside came to stand for the past and people's origins.

RELIGION

For Macedonian citizens, the modern city negates the rural ties that are assigned as a means of identification. Instead the urban/rural divide is translated into an ethnic divide with the assignation of modern characteristics to Slav-Macedonians and of backward ones to Albanians. As the new Macedonia has to define its future once more, it has to solve the puzzle of this dichotomy. The most obvious solution for a state that has supposedly[17] thrust off its socialist past would be to present religion as a new common denominator of unity. However, this creates another level of friction: the Christian/Muslim divide. In recent years, Skopje has been confronted with increasing numbers of state-organised Christian rituals, which establish the Christian Macedonians as inheritors of the Macedonian state. As Poulton writes (1995: 182), "It is hard to assess accurately the strength of Orthodox religious belief in Macedonia but it appears that many see the church as the key to nationhood. This role of the church as an essential constitution inevitably alienates the predominantly Muslim Albanians."

In 1990, the Soborna Crkva Kliment Ohridsko, the Congregated Church of Kliment of Ohrid, was completed. Throughout the summer of 1990, the date of its completion was rumoured to be the day that an earthquake would strike Skopje for a second time. People were in panic. It seems now that this rumour had more truth in it than many people realised at that time, at least metaphorically. The Macedonian emphasis on Christianity excludes the growing Albanian population from the state. An earthquake in response to the completion of a monumental Eastern Orthodox church may, then, be an appropriate metaphor for a civil war occasioned by this exclusion, as one informant suggested. It is also significant that in the struggle for an Albanian University in Tetovo,[18] the University of Kiril and Methodi became the University of the Saints Kiril & Methodi. In this respect, it is the Macedonian people, although they may not agree explicitly with this analysis, who are taking on the role of the cultural aggressor, imposing state symbols derived from a Christian source. Thus, it seems, Christianity might be desired in the formation of nationhood and identity, but it has not succeeded in this role until now. Under *millet* organisation by the Ottoman Empire, religion stood as the defining factor of a group. When people say they are Orthodox, they are claiming not the religion so much as their Slav origin.[19] As I indicated earlier, the people of Macedonia, largely because of their historical and geographical position, felt affiliated to a specific soil, to a specific village, and to the church of a place, but never to an outer whole. This complex series of attachments we find again in the group of my informants, who identify themselves as belonging to a specific place. There is another aspect of religion that becomes vital.

Christianity does not seem compatible with the young people's struggle in the urban environment of Skopje to become a modern, Western people because they see the Western world as a secular world.[20] Neither the idea of the past nor religion is an easy point of departure for creating a "Macedonian identity."

CONCLUSION

When one examines the different threats or obstacles to the identity formation of the youth of Skopje, several things become apparent:

1. It appears that people are not decisive about their own identity and cultural heritage, which is seen in how young people in Skopje seem to be caught between worlds. This indecision suggests that the state has not succeeded, or perhaps does not wish to succeed, in forging a distinct Macedonian culture into an object of consumption. A distinct Macedonian culture appears to lack attractiveness to its citizens. Even though people might be presenting themselves as "proudly peasant," belonging to the Balkan heritage is equated with backwardness and not in itself a condition to which one aspires.
2. The rural past is seen as lived out today by the Albanian population, and both are perceived negatively. In this situation, people acquire a cultural heritage, but that is a negation of everything one wants to be. In this respect, the state cannot use "culture" as an object of "worship" nor as a fundamental legitimisation of its power. Further, given the external challenges to its identity as a state, the Republic of Macedonia must attempt to achieve an understanding with the Albanian population for the state's own survival because the republic needs strong internal co-operation and integration in opposition to external forces. The very real threat of civil war still exists, and seems to have been avoided only because of the negative example of Bosnia and the careful response of some intelligent moderate politicians to potentially similar situations being repeated in Skopje.
3. Religion is not seen as belief but as a denominator of identity. My informants and the church officials with whom I spoke represented religion as connected to oppression, in alignment with socialist ideology and because religious power in the past (through, for instance, teaching positions in the schools) was held by Serbian, Greek, or Bulgarian authorities. As a denominator of identity, religion has become a powerful instrument to distinguish between "us" and "them": Orthodox Christianity versus Muslim identity, but also non-belief versus belief or modern versus peasant and "West" versus "Balkan."

Skopje itself is a powerful metaphor for the experiences of my friends and informants after the disintegration of Yugoslavia. The country's boundaries were redefined, and citizens started to look at Skopje and not Belgrade for leadership. Essentially, what Skopje represents is differentiation between various strands of "identity" that together weave the new country. Within it, this tapestry carries the thread for the future of Macedonia. Skopje represents the differences of the country in an ongoing dialogue between urban and rural, young and old, Slav and Albanian, secular modernity and Christian Orthodoxy or Islamic religiosity—between the "West" and the "Balkan." Within this dialog, space is recreated, space that mediates memories of other places. Some of this space is created through the memory of a rural past, some of it through media and foreign aid agencies. By assuming the role of dependent recipient in the global aid economy, Macedonia creates the idiom of the "West" as a commodity just as the socialist state created folklore as a uniting commodity. The rhetoric of the non-civilised Balkan versus "The West" collapses time and space such that my young informants see themselves today on the margins of a meaningful universe as consumers of an externally generated material modernity.[21] The landscape of Skopje, however, in the midst of "conflicting spaces" also creates a relationship between the life lived and an imagined existence of what Macedonia could be. As such, the landscape of Skopje describes a process, its transformation, and its power to transform.

NOTES

1. Numbers can be drawn from the 1994 census.
2. For a thorough discussion on the Macedonian-Albanian conflict, see Poulton 1995.
3. Although *Skopjanici* means the inhabitants of the city of Skopje, Albanians and Roma living in Skopje are always called Albanians or Roma.
4. How these attributes were assigned is outside of this discussion and of my understanding of the folk tradition of these places.
5. See *Being Muslim the Bosnian Way* (Bringa 1995: 25) for a further discussion on the Yugoslavian understanding of the terms *narod, nacionalnost,* and *narodnost.*
6. My informants never expressed a feeling of estrangement. Many people are indeed proud of their city, especially of its nightlife and "café culture." However, people did apologise for the "Balkan feature of their city." They did so assuming that I held "Western European standards," and they related the term "Balkan" mainly to the dirt and destruction throughout the city not to the city's overall appearance. The fact of my informants' estrangement I have extracted from their comments about their origin.
7. Food is a resource that is still heavily stressed in social interaction.
8. Gold still plays a major role in the family relationship. It is gold that expresses kinship—the presentation of gold emphasises kinship ties.

9. This migration was enforced by Tito's destruction of the "rural bourgeoisie," which left many villages deserted, villages then taken over by the Albanian population. This migration was due also to an influx of Albanians from Kosovo. As Slav Macedonians claim rural origin, they are very bitter that the countryside, in many places, has been taken over by the Albanian population.
10. See Alan R.H. Baker, "Introduction: On Ideology and Landscape" in *Ideology and Landscape in Historical Perspective*, ed. Alan R.H. Baker and Gideon Biger (Cambridge: Cambridge University Press, 1992), 4.
11. I was always told that Japanese architects designed these houses and, therefore, they are earthquake-proof.
12. See Alan R.H. Baker, "Introduction: On Ideology and Landscape" in *Ideology and Landscape in Historical Perspective*, ed. Alan R.H. Baker and Gideon Biger (Cambridge: Cambridge University Press, 1992), 10.
13. Interestingly, as I mentioned, the Muslim signifiers stood untouched. In the discourse between the Macedonian and Albanian population, the survival of historic Turkish architecture becomes vital as the Albanian population does not have a problem with its own identification of what it means to be Albanian.
14. See G.W. McDonogh and R. Rotenberg, eds., *The Cultural Meaning of Urban Space* (Westport, CT: Bergin & Garvey, 1993), xii-xiii.
15. See Andreev and Jakimovski 1993: 124. The government today, trying to privatise the housing sector in the hopes that people will take better care of their "own property," offers its citizens the use of frozen foreign currency savings to buy their own flats.
16. In the 1994 election, the division of the city was more apparent. The inner (South) city voted predominantly for the "Reform-Socialists" under Kiro Gligorov. VMRO gained strong support in "Slav" areas in the North, which also house rising numbers of Albanians.
17. I say supposedly here, as it is not clear if Macedonia really has set its socialist past aside. Kiro Gligorov is an "old communist" and so are many others. What is now the Social Democratic Union was once the League of Communists of Macedonia. Many people have called the system neo-communist rather than democratic, although the government includes many young professionals who did not belong to the socialist party, and certain democratic freedoms, such as freedom of speech, are guaranteed, as far as I can tell. However, as Poulton (1995: 206) points out, "there were few dissidents in Macedonia under the old Communist system, which is perhaps an indication of the lack of any democratic culture in the Western sense."
18. An Albanian was shot dead by the police in Tetovo while demonstrating. However, far from belittling this, I want to mention that the man shot was a Kosovo Albanian. The Kosovo Liberation Army lately have started playing a major role in militarising the Albanian population of Macedonia.
19. There is a problem of Slav Macedonians who are Muslim that I am not clear about. I have only very limited information on this, but the Muslim "Macedonians," as they call themselves in opposition to Albanian and Turks, do seem to identify themselves as "Slav." The Macedonians do not equate them with the Albanian

population, and the two groups mix without evident difficulties. Only in regards to inter-religious marriage, even if both partners were atheists, would religious affiliation play a role. "Two religions are not to belong under one blanket" is a Macedonian proverb.

20. Information about abortion in my own country was received with astonishment. When told that a majority in my country seek to make abortion illegal on religious grounds, my Macedonian friends cried, "We thought you were a civilised country."
21. See Fabian 1983.

Chapter Three

The Disintegration of Yugoslavia: Gender—Experienced by Three Generations

Concurrent with my argument so far, the lack of an analysis of gender relations in Macedonia makes an assessment of the impact of Western images of these relations on the actual conflict between Slavic Macedonians and Albanians impossible. The constructs of the backward Albanian and the modern European are gendered and draw heavily from images of gender relations. The Balkans are described as the place where wives still have to wash the feet of their husbands, and Europe is seen as the place where women become innovative engineers.

One should not overlook Marxist theory, which played with the possibility of an egalitarian social order erasing the roots of female subordination. Certainly, this idea once played an important role in Macedonia, especially for the young Macedonian women who entered the engineering faculty in Yugoslavia. A central question to my understanding of the social change that was going on in Macedonia was whether my young informants assumed the existence of a prior egalitarian social order. It was a question often hotly discussed among us. With Friedl and Lamphere, I argue that Macedonian women in the past, even though they appeared to be powerless and without formal power or authority, were not without individual power.[1] In my interviews, elderly village women in Macedonia described their subordination to their husbands as something arising from a specific economic and historical position rather than as something innate. And it was during these interviews that the elderly women would tell me sometimes about the violent and erratic behaviour of their husbands during their youth, when exactly the same spouse would enter and kindly offer us some mountain tea he had collected and brewed with care.

GENDER

Interestingly, when discussing their relations with boyfriends or husbands, the city women whom I interviewed described the status difference between the sexes not as inequality in the household or in the workplace, but from within the difference between Macedonian and Albanian women. This result underscores my argument and shifts the analytic focus from simple dichotomies to a far more complex system of images and identifications.

In the new political order, social relations have become critically important although political theoreticians in Macedonia do not recognise this. Politics is represented as gender neutral. There are more urgent issues than women's rights, a government official stressed in one of my interviews. Consequently, political discourse centres on privatisation and inter-ethnic conflict and does not address issues such as the changing values in society. Despite, or perhaps because of this neglect, certain socialist achievements around gender equality remain legally untouched, providing a stark contrast to the situation in other post-socialist countries that chose to overthrow their socialist inheritance. Abortion is still legal, and there are no calls for it to be restricted or prohibited. The proposed changes to working hours, which accompanied the privatisation meant to "Europeanise" Macedonia, have been successfully resisted, and labour laws continue to allow men and women to finish work at three or four o'clock and spend time with their families afterwards.[2] Macedonia's socialist past is not resented; indeed, it is sometimes glorified and is very much present. However, another world has been introduced, a world that is supposed to pave the way to a different Macedonia, one with a distinct identity that is a blend of the old and the new.

Strathern (1980) in her article "No Nature, No Culture" stresses that gender is a symbolic system, that the people of Mount Hagen in Papua New Guinea do not assume that one gender category is dominant over another, and that dominance and subordination are Western concepts inappropriate in a non-Western context. Though I do not claim Macedonia to be non-Western, I do argue that, in Macedonia as in Mount Hagen, at a time when the Marxist definition of gender relations was being questioned and re-negotiated, gender relations were created as symbolic systems with specific attributes, attributes that were not men/women, but Albanian man/Macedonian man, Albanian woman/Macedonian woman. I also argue that the definition of the Albanian man in relation to the Albanian woman was fixed in time and place (that is, in the past and in the Balkans) but that the definition of the Macedonian man/woman relationship was being re-negotiated. Therefore, just as in Mount Hagen, in Macedonia particular women and men could transcend gender boundaries through specific actions. These actions, I claim, were put through in a "Western European" framework, or in what the images that were transmitted and received promised was "Western European."

Gender relations, then, became a symbolic system that referred to ideal behaviour. Analysing gender relations in Macedonia as a symbolic construction that is based on images of "European" types of action makes it possible to analyse the relationship between the bodies of my Macedonian friends—the symbolic representations of what it means to be a woman or a man, which they seek in definitions of "Albanian" and "European"— and the actual behaviour of individual women and men. To facilitate the understanding of these relationships, I argue that the value system being introduced by the "Non-Eastern European," by the non-post-socialist world, defines the values of being a man and a woman in Macedonia. Consequently, in talking about gender roles in Macedonia, we are actually talking about a value system that pertains to Western Europe. Gender becomes a symbolic system, which provides images for the ordering of social persons in relation to each other and to the social system as a whole. In *Contested Identities,* Loizos and Papataxiarchis suggest just that. In discussing the cultural construction of masculinity as a missing aspect of gender, their book argues that gender should not be perceived just in the context of kinship but that, indeed, it often stands in conflict with kinship. Cornwall and Lindisfarne in *Dislocating Masculinity* also emphasise the greater framework of gender, which reaches beyond kinship and beyond the category of "woman." In this respect, I want to emphasise that Macedonian gender relations are not just used to talk about differences between men and women, but about differences between Albanians and Macedonians, between the Balkan and the European, and between the past and the future.

GENDER ROLES AND RELATIONS

In 1995, I had a conversation with Dragica, a mother of three who lived with her family in a high-rise apartment building in the centre of Skopje. In her small living room, sipping Turkish coffee and eating very sweet cake, we reflected on her views about the changes for women over the last fifty years. She was adamant that the life of Macedonian women was very different from what she thought the life of women in the West was. She expressed this difference in her concern that women in socialist Macedonia had a far harsher life because they had to work and could not stay at home and take care of the children. If a woman in Macedonia decided to stay at home because, for example, she needed to care for a child with special needs, she would find no support from the state. Dragica explained that women in the West had something she called social security even when staying at home, something Macedonian women did not possess, and this lack was why all Macedonian women had to work.

> We have to work, to finish some school, to have good job, to have social security. We have to be successful at work in order to be paid well. We have to be a good mother because our husband does not help too much with the work at home. Macedonian man, they go to work and then come home and expect the food on the table; that's all they do.

Dragica felt that the younger people of today thought more about how to have a good marriage in which both partners were happy and satisfied with their lives. She felt that the young people had learned this from their travels abroad, visiting Germany, and from what they watched on television. She described her life under socialism as very monotonous. "I was working in an office, then I would come home, and I had to feed the children, clean the house, wash, cook, do the laundry. In our lives, it's very, very difficult." Even the lives of her younger female colleagues she describes as still very difficult but changing.

> I have colleagues who are young. In this time, wintertime, they have to wake up at 5:00 in the morning, they have to wake up the children and then bring [them] to the kindergarten, and then they run to the buses that often do not come. In the bus, it's so crowded with all these people inside who all try to come to the office on time. If you come late, the boss makes problem. Our working time usually starts about 7:00.

Dragica explained that if a woman could choose to stay at home, it would be best for everybody. The children would be happy and not nervous. They could sleep in until daylight. For children to wake up at 5:00 in the morning and be put into clothes and blankets, for women to have to run to the kindergarten and then to the bus station and after work to the stores and back to the kindergarten to pick up the children, all this was exhausting; Dragica clearly felt that when women stayed at home, it was far more advantageous for everybody. That this was not possible she explained as a fault of the system that paid men and women the same, making it impossible for a family to live on one salary. And she felt that the parliament was mistaken not to change this wage scheme; after all, Macedonia was its own nation now, so why did the government not think of making life easier for people instead of more difficult? I was interested in how she would support her point that the parliament was making life more difficult for people, and Dragica explained that it was the new market economy that was driving everybody to make more and more money instead of helping to make life more easy and relaxed.

Through these paradigms of hers—"nervous" and "relaxed"—Dragica reflected on the changes that were going on in Macedonia at that time. Nervousness was caused by the old socialist regime, which neglected to see what is described commonly as the "double burden" of women in the socialist system.[3] However, it was also caused by the newly introduced market regime, which neglected family values once again. In contrast, what Dragica termed "relaxed" was what she saw as being "Western European." This relaxed world was the world presented by Western television, a world where women stayed at home and cared for their children and the husband worked and came home to relax and enjoy some family time. According to Dragica, this world was currently impossible in Macedonia:

> No man would understand this and say, "You have a child, you go home, and I work longer for you." No, [a] man would say, "We are the same, so you stay when I stay." No man would understand that you had to stay at home because your child would be sick, but instead they cut your pay.

Dragica's paradigm of "relaxed" or happiness was measured by the happiness of her children. Whatever would make the children happy was good, ultimately. So in her discussion on abortion, she insisted that abortion must be legal, because, if a woman were forced to have a child she did not want, the child would be a very unhappy child. It was through her children that Dragica felt she fulfilled herself, so she reflected on the changes around her through them as well. The roles that women and men should have were centred on the question of who could take the best care of the children. And it was through their children that women had control over things; if women could guarantee the happiness of their children, Macedonia as an independent state could survive. In this respect, reflecting on the difference between Albanian and Macedonian women, Dragica was very clear that there was no difference between Albanian and Macedonian women because they were foremost mothers. In some respects, it was through this notion that her generation still put forward the Yugoslav ideology of unity and brotherhood (or motherhood). The younger generation, however, when reflecting on gender relations, drew new conclusions.

Sonja, twenty-seven years old at the time of the interview, is very active in a Board on inter-ethnic understanding, the Macedonian Centre for International Corporations. More specifically, she works in the Centre for Ethnic Relations. She is also a member of the Executive Committee of the Macedonian Council of the European Movement and of the Peace Movement. The Board on inter-ethnic understanding dealt with humanitarian issues and was financed by the Netherlands, which gave it the mandate of giving relief to people in need. For example, Sonja was active in establishing night schools for Rom in Skopje and in finding mattresses and beds for people who needed them. This Board was also concerned with projects that would create employment for minority women. According to Dragica, this last initiative was, of course, regressive instead of progressive. However, according to Sonja (and the Dutch mandate), minority women's employment brought progress to backward people. I asked Sonja if these projects included Albanian women, but, in her understanding, Albanian women were beyond the means of rescue:

> Albanian men are going abroad to Switzerland and Germany. So they are manual workers or skilled workers, depending if they finished school here. So they just send money to Macedonia, and women are staying home and taking care of the children. As such, the men cover the costs of living here, so women are not really motivated to go and work, and that's a problem. But on the other hand, for example, Rom women are very dominant in their family because they're the ones who bring money in, and men are the ones who are sitting around. These are stereotypes, but I'm telling you about the general outlook on our situation.

When women try to achieve a state of happiness in Dragica's terms, Sonja points out that the foremost problem for the Macedonian State is that it does not have money and, consequently, needs to rely on humanitarian aid or foreign humanitarian organisations.

The other problem Sonja saw was the lack of what she termed "progress" in the minority groups. "Backwardness" finds its expression in the high birth-rate among the Muslim and Rom populations, and, in turn, leads to the declining economic prosperity of the Macedonian state: "... it brings poverty most of all. It's a tradition, but it's one thing when you have two children and another thing when you have five or six of them." In order to bring progress to ethnic minorities and decrease the poverty of the Macedonian state, Sonja advocated what she called the "liberation of women." If Rom and Albanian women could be integrated into the work force, Roma and Albanians could be led into modernity. At the same time, Albanian women would have fewer children because they would want to fulfil their individuality and not live only for their husbands like a "Balkan woman," as Sonja stated. Albanian women would have access to abortion, and the ethnic conflict in Macedonia would be solved. Sonja's concept of a newly changed Macedonian society was, in contrast to Dragica's, centred on the idea of women, especially Albanian women, being more integrated into the economic and political process of a new independent Macedonia: "women are getting better positions now, and you can find even single mothers, marriages that are not legal. The trend is coming in from the West." When I asked Sonja how this change was brought about she insisted that it was not a new development, that socialism had paved the way for women in Macedonia. And, in contrast to Dragica, a woman the age of her mother, Sonja emphasised that the integration of women in the workforce was the specific factor that brought progress to Yugoslavia. Today, progress would come from the integration of Albanian women into the workforce, which, ultimately, would bring Macedonia into proximity with Western Europe.

> Maybe if I speak of my mother, I think about a woman who's an engineer, active, and everything. But when I speak of my grandmother, I always think of somebody "uneducated," a "villager"—you know, you met my grandmother. In her family, her brother was an economist, director of a bank. The other [brother] was working in the hospital. The women in her family, however, only learned how to write, how to write a letter and that's all.

In Macedonia, then, young people like Sonja, defining progress in socialist terms, draw parallels between the progress needed within the Albanian population and its specific gender relations and the progress that socialism brought to the rural life of their grandparents and to their grandparents' gender relations.

In retrospect, the voice of the grandparents' generation becomes interesting. One of the most vivid conversations I had was with Baba Vesna in a suburb of Skopje. Vesna was born in a Turkish town in 1910. The Turks occupied her town until 1912, and then it was occupied by the Serbs until 1915. From 1915 to 1918, the Bulgarians were in control. During this time, some rumoured that the United States would take the town as its protectorate, but this did not occur. After World War II, Vesna's town belonged to Yugoslavia, and today Baba Vesna lives in Macedonia. With dry humour she commented, "Whoever came, the Serbs came, then you were a Serb. Bulgarians came, then you were a Bulgarian, today we are what?"

When Vesna was thirteen, her mother died and Vesna's older sister started taking care of her father "who was incapable of doing things." In 1928, Vesna learned to sew and was, at that point, the only one in the family earning money. In 1930, at the age of twenty, Vesna married a priest. According to Vesna, her father was very upset with her choice and tried to dissuade her, but, because Vesna was supporting the family at the time, she claims that her father did not have the right to tell her anything. After her wedding, Vesna stayed at home and raised three children. Nevertheless, when she first met her husband, he was a young priest without money, so she sent him 100 dinars every month, money she earned by sewing. In return, Vesna told me, he sent her a postcard every day, with loving things written on each one. Vesna confessed that she had burned this correspondence when her daughter started to read. She did regret this today, but then she was worried that her children might read something indecent.

Vensa's first daughter became a mathematics teacher. Her son did a Ph.D. in France, and her second daughter did a Masters in chemistry. In 1950, the socialist authorities imprisoned Vesna and her husband. The socialist regime was starting to collectivise land, and people had accused her husband of initiating a rebellion. They spent several months in prison but were released due to insufficient evidence. Of course, I wondered whether Vesna disliked Tito because of this incident; diplomatically, she told me that she had lived a good life under Tito and that she was able to travel a lot with her husband and some pensioner groups. She went to Serbia, Greece, Paris, Naples, Germany, and Hungary; she lived in London, visited Jerusalem, and saw Istanbul, Syria, and Beirut.

Certainly, Vesna's life cannot be seen as typical; however, it was also not unusual. What was unusual, though, was that the younger generation insisted on the terrible lot of their grandmothers, who had not known love and happiness and who, according to their granddaughters, had lived for only husband and children. Certainly Vesna's life offers a very different perspective, a perspective that conflicts with the social discourse on progress and modernity. Her life cannot be paraphrased as a life of oppression that was overcome by socialism, as a way of life that today has to be overcome by another kind of progress, the progress away from the "Balkan" and toward the "European." This particular image of progress is endorsed not only by young urban Macedonians and politicians but also by Europe itself.

SEXUALITY

In a time of social change, where old is replaced by new but where new is defined through the familiar, through things that have been passed on and now define a new future, through inherited images and notions of a future that are created anew—in such times, how do people understand themselves as subjects, or how are they made to be social subjects? Ong's work on Malaysian factory workers (1987) deals with this question when considering how young women are made into good factory workers in Malaysia, and I will operationalise some of the points she makes in her argument. Starting off with Foucault's understanding that all human beings are placed by social life into relations of production, relations of signification and relations of power, I argue that, in relating to the image of the Balkan and to the image of the past and to the image of Albanian women, the young Slav women in Skopje *did* change what Foucault calls "techniques of domination" (see Foucault 1995). Consequently, these young Slav women not only changed their ways of talking about things, but, ultimately, their ways of doing things—mundane things like shopping, eating, and exercising. From these changes, I argue, one can read how my informants understand themselves as selves in the present. As Foucault argues in the *History of Sexuality*, power and knowledge are fused together in "modes of domination." These "modes of domination" can be found anywhere, even in such things as attitudes to sexuality and the practice of sexuality itself.

Sexuality was a frequently discussed issue among the young women in Skopje (and among the old women in the villages, I discovered to my surprise).[4] Mostly these discussions concerned how and where to have sex: where because it was impossible to have it at home while sharing a room with your siblings and having your parents right next door. The how to have sex centred on the question of using condoms. This question was, in itself, a highly modern, Western discourse. Back in the eighties when we discussed prostitution, my young friends were adamant that there was no prostitution in Yugoslavia. The sexual and economic exploitation of women in Tito's Yugoslavia was not possible, certainly not in Macedonia. When returning to Macedonia in 1991, I was welcomed with some sheepish looks on the face of one of my friends, the person with whom I had fought about the existence of the oldest trade in history. She admitted, with some giggles, that she had found out recently that she had lived right next to the biggest brothel in Skopje. Now, this brothel had been closed and turned into a hotel for foreigners. That did not close the discussion on prostitution, however. More and more stories emerged about UNPROFOR soldiers and ruthless businessmen who wandered into the old part of town for the services of younger and younger women. These women, though waiting for their customers in the Albanian part of town, were always defined as Slav women; the thought that Albanian women would be prostitutes was unimaginable. Bound by religion and tradition, Albanian women were also guarded by their families, and, as prostitutes, they would have brought unimaginable shame on these families. At least that was the image. Interestingly enough, Rom women were never mentioned in respect to prostitution either, even though they were portrayed as easy-going and dominant.

It is important that the discourse on prostitution, sexuality, and pornography centred on what my young informants thought of as Macedonian women. Macedonian women were living in a world of change, whereas Albanian women, Rom women, and village women were portrayed as living in a stagnant world of tradition and past values. Even though the prostitution of Macedonian women, mostly young girls, was not endorsed by the broad public, and the girls were pitied and the clients chastised, prostitution as a whole (and pornography also) was read as sign of change and modernity. After all, everybody had heard of Amsterdam.

Like prostitution, the discourse on AIDS was read as Western too. The summer of 1992 was defined by people's inability to travel to Greece for holidays, as they had every other year, because of the embargo. They could not travel to the Croatian coast either, for obvious reasons. Many young people from Skopje decided to travel to Bulgaria instead, and the young men came back with tales of amazing sexual exploits. One particular man boasted about having slept with a different woman each night for two weeks, one night even sleeping with first the mother and then her daughter. The reason that women in Bulgaria seemed to have been enchanted with Macedonian men was due to the greater prosperity of Macedonians at the time. It was said that the greatest hope of one of these women was that a Macedonian man would fall in love with her and take her to Macedonia. This story tells us several things. First, it portrays Macedonian men as superior to Bulgaria as a whole. Like "Westerners," they came in and "appropriated" local women, just as UNPROFOR soldiers sought out Macedonian girls. The story also tells us something about the sexual talents of Macedonian men, something that both the women and men present during those conversations were eager to underline for the Western anthropologist. I do not need to draw the parallel between sexual strength and power. From these stories, however, a heated debate arose, a debate that centred on whether the men had used condoms. "Of course we did not use condoms," the men said, which was met with an outcry from the young women present. "Had they never heard about AIDS?" was the women's reaction.

The denial of the possibility of contracting AIDS, then, parallels the denial of the existence of prostitution in Yugoslavia. In some ways, the discourse around AIDS was presented as a Western and modern discourse that elevated the women present to Western European status and left the men accused with being typically "Balkan"; they certainly would never score in this way with Macedonian women, so they had to go to Bulgaria. Macedonian women were aware of the consequence of unprotected sex, that is, an unhealthy body, and they took control and insisted on men using condoms. At least, this is what I was led to believe. However, when many of the young women I talked to became sexually active, the reality was quite different.

In order to have sex, one first needed a boyfriend with a car. The car was needed as the locus of the deed. A couple would drive up Vodno maybe, or sometime be adventurous and make out in the parking lot. Condoms were never used. I heard different reasons for this. The condoms that one could buy were not of good quality: they were old, they were smelly. However, women told me that the real reason no condoms were used was that their partners were convinced that condoms took away a man's strength. This rationalisation was presented as a silly argument, men being somehow less enlightened than my friends, but it did not change the fact that my friends were very happy not to use condoms. Intrigued by this, and by the fact that condoms and sex make good subjects of conversation, I pushed this issue further. For some of my friends, and I am certainly not claiming this to be true for everybody, the non-use of condoms was primarily the signal of a partner's commitment. If it should happen that the woman became pregnant, a wedding was inevitable. Having a husband with a car and maybe even his own room or apartment secured an escape from the crowed quarters at home and promised an independent life and some sort of control. The issue of AIDS was brushed away with the conviction that certainly no Macedonian man had contracted it. Not using condoms, therefore, confirmed the integrity of their bodies, their modern bodies. Illness and death were things of the past, and a Western lifestyle, as promised in television advertising, would guarantee an impermeable body. In effect, the body of my young friends, created as a "European" body stood in opposition to the deceased body of the past. Talking about AIDS was a modern, European thing to do; practising unprotected sex was a confirmation of the control that one exercised as a modern European.

SEXUAL DIVISION OF BODY USE

Whereas men are boisterous, wide-armed, and loud, women in Skopje are graceful and cultured. Often a man is described or describes himself as a "Balkan boy" whereas young women see themselves as "European" and refined. To represent this difference, I want to recall one particular event that illustrates nicely the sexual division of body use.

One warm evening, when a warm breeze came up from the Vardar and drifted along past crowded cafés that had more people standing outside them than in them, I sat on some stairs enjoying the vibrant life and the chit-chat around me. Suddenly, a charming Macedonian man fell before me on his knees reciting to me Shakespeare's *The Taming of the Shrew.* Despite the questionable speech he chose, I was certainly charmed by this act, and neglected the fact that the young man was obviously intoxicated. This is how I met Val, who was himself only visiting his hometown for the summer, as he was studying abroad. His friends knew my friends, and, before we knew it, we would run into each other in the cafés on many evenings and often decide to continue our nights together going from café to café. Sometimes our choice of café required the use of a car, and we would pile into the few available Ladas and Wartburgs.

As it happened, one evening I was separated from my friends and piled into the Lada of a friend of Val's with about five other men. Val's friend turned out to be called "crazy Igor," a fact which, had I known it initially, would have given me great cause for concern. But I was peacefully sitting in the Lada while Val and his friends tried to squeeze in without squeezing me too much. Even my attempt to buckle in was tolerated with mild laughter. We went on a memorable journey, driving through the streets of Skopje at 100 kilometres per hour. As if this were not frightening enough, we were soon engaged in a car race with another crowded Lada, which was trying to take the lead from crazy Igor, who tried to prevent this catastrophe. While we are racing, one of my companions had the great idea of opening our car door and crashing it into the other car, which was not welcomed peacefully. The other car eventually took the lead and drove right in front of us, causing crazy Igor to stand on the brakes and me to praise the foresight that had me wearing a seatbelt. Then, all the men jumped out of the car. (This seems to be the reason that men in Macedonia do not wear seat belts—just a thought). Val, torn between his duties as a Macedonian man and a European gentlemen, handed me his can of coke with the words "I have to help my friends" and jumped out. I heard bottlenecks break and was expecting the worst when the driver of the pursuing car discovered he was a close friend of one of my companions. Before I could take another breath, the men in the competing cars embraced one another, giving and receiving big kisses on cheeks and lips, and I was relieved to hear great roars of laughter. Because I have taken great pleasure in re-telling this story over the years, people have had ample opportunity to comment, with great pride, on how "Balkan" this event had been.

Walking in Skopje with my friends on a shopping spree, however, I often encountered Albanian women, veiled and walking behind their husbands or guarded by their sons. Continuously, my friends would comment on how "Balkan" this sight was and that their grandmothers, not long ago, had lived very similar lives but were now advanced whereas the Albanians, bound by religion and tradition, were held back in the past. From these comments I concluded that the definition of Balkan could be twofold: male and proud or female and backward. Women, therefore, especially young women in Skopje came to signify modernity and change, came to signify Europe. This modernity was lived and practised by my young friends who were very proud of the fact that they, as female engineers, belonged to an elite of European women.

NOTES

1. See Friedl 1986 and Rosaldo 1974. Also see Thiessen 1999a and 1999b.
2. In general, Macedonians are unaware of discussions in Europe or America concerning the problem of long working hours.
3. See Einhorn 1993.
4. See Thiessen 1999a and 1999b for more on villagers' sexuality.

Chapter Four

Getting Along

During the times that tore Yugoslavia apart and swept the world with images of atrocities and stories of friends and neighbours killing each other, the idea of "getting along" seemed aeons away from everyday reality. Macedonia, fortunately, has been able to stay out of the civil war in Yugoslavia and also to prevent civil war within its own borders, so far. Understanding what it means to live in these times of uncertainty and change requires consideration of how daily life proceeds in Skopje while the former Yugoslavia is drowning.

On the surface, life in Skopje seems the same as usual, but maybe it is not. The circumstances of the young female engineering graduates, who after a worry-free adolescence in Yugoslavia are now entering the work force and adult life, illustrate clearly the trials of daily life. Vital to this discussion is the difference between what my informants term "Western" and "Balkan." The "Balkan" they see as the cause of the Yugoslav civil war and the "Western," by definition, as anything in direct opposition to the "Balkan." Tone Bringa discusses the same notion in her book *Being Muslim the Bosnian Way* in terms of being civilised and uncivilised, *kulturni* and *ne kulturni*, cultured and uncultured. In Skopje, these terms, *kulturni* and *ne kulturni*, existed on the same local and global scale in the form of "Western" and "Balkan." Whereas Bringa's discussion of these terms focuses on the difference between the city and the village, in Skopje, at the time of my research, the city itself and its constituents were categorized as "Balkan" or "Western," and these terms established a scale of culturedness with "Western" at the top descending to "Balkan." This culturedness was assessed in relation to people's behaviour in social interactions with others (Bringa 1995: 58). Therefore, creating a partial picture of daily life in Skopje reveals elements of social change within Macedonian society.

I do not claim to represent "the truth." "Truth" cannot depict life with its ambiguity and shifting perceptions, life as it is lived by people. But one of the "truths" I was told in most of the conversations of my fieldwork, was the "truth" about the difference between the "Balkan" and the "Western." For instance, Goran identifies his specific understanding of the "Western" work ethic as the precise and organised assignment of tasks to the workforce and the punctual and early start of the workday, and he contrasts this ethic with the "Balkan" attitude of his supervisor:

> When I got the job as music editor at Zatex, I did not really know what I was supposed to do, and I still do not know. I do a little bit of this, a little bit of that and, basically, I can do whatever I feel like doing. The boss is never there anyway, and he does not care. When I started, I always came at 8 in the morning. One day my supervisor came and told me I was making others feel bad. Why couldn't I come a bit later, like 11 am? Now I come around this time. I talk to my colleagues and then we go to the café. Only when the boss is in do we not want to be seen sitting in the café.

Although (or perhaps because) Macedonia lies in the heart of the Balkans, in Skopje to be from the Balkans is seen in a negative light. However, when Macedonia was a republic within Yugoslavia, its physical boundaries were extended towards Northern Europe or "the West." With the disintegration of Yugoslavia, those boundaries have been redefined, and Macedonia has been thrust back into the Balkans. My informants feel that this has had very real effects on their lives, the greatest of which has been the alteration of what they expect in their adult lives. Concerning work, they were, as they said, highly ambitious and oriented towards "the West." They were heading for a "career" and wanted to work, to improve their knowledge in their respective fields, and to do something useful. They also wanted to earn money, but this was secondary. What they got in independent Macedonia was, in contrast with their "Yugoslav" expectations, not a career, but a job, and a poorly paid, dead-end one at that. For many, their jobs did not offer the possibility of promotion to more rewarding and challenging positions. Furthermore, their way into work was facilitated by their parents' "connections," thus creating a new dependence. My informants called this situation "Balkan."

For many of my informants, a low wage was not an issue in terms of their contribution to a household; they used what they earned as pocket money. Some of the more entrepreneurial tried to do some consulting work that was directly connected to NGOs and to foreign investment in Macedonia. This work promised, eventually, more money because most of the enterprises hiring consultants were subsidised by foreign capital and expertise. However, to be involved solely with these enterprises was considered too insecure by most parents and therefore a relatively poor job choice. The perceived inferiority of these endeavours was based, I believe, on two factors:

1. Parents preferred traditional, secure employment in state organisations;
2. Foreigners, the feeling was, could pull out of Macedonia at any time because of lack of interest, economic factors, or war.

My informants are being initiated into their new role of "making a living." They are trying to perform in a work culture, however, that neither gives them the experience of "really making a living" nor develops them for roles they would like. In this context, what does work mean to my informants and, if they see themselves working, how do they respond to the specific limitations their new society imposes on them? By relating three cases, I give a picture of what has made the working life of my informants so different from that of their parents. This depiction also outlines what Yugoslavia may have been able to offer these young workers a few years ago. When people considered the situation in Western Europe after the world wars, they seem to have believed that life could not get worse, only better. My friends, on the other hand, carry in their hearts the thought that things can only get worse or, at best, remain unchanged.

THE CHANGE

Before considering how my informants experience the challenges their changing society poses for them, an account of the world around them is necessary. This information is based largely on an informal interview held in May 1995 with Nane Ruzin, Member of the Parliament of the Macedonian Assembly, who sums up the current situation in Macedonia with such insight that his thoughts mirror those of most Macedonians I have met. For many people, the dissolution of Yugoslavia has meant rethinking what they believed during Yugoslavian times and discovering what to believe today. At the time of my research, what Macedonians could believe in was not clearly defined for most people, including for Minister Ruzin.

Upholding socialist values was becoming increasingly difficult as the all-encompassing framework of Yugoslavia disappeared and nationalist identity was defined relative to the contrast between "Macedonians" and "Albanians." The redefinition of the state boundaries, which no longer reached toward Austria or Italy, Macedonians greeted with a partial glorification of what they perceived as "Western," that is, of what had been close to the physical boundaries of Yugoslavia and accessible to them through travel. An important Macedonian feature of this glorification was that the legacy of Yugoslavia was not forgotten and was also glorified. Such perceptions melted into an interesting amalgamation of a Yugoslav past with some Western ideals, a composite that people contrasted to a Balkan identity. People feared that they would lose the connection they had with the West and regress to being Balkan again, rather like the Albanian minority is perceived being in the process of doing. The economy of Macedonia is, despite financial aid, battered since it relied heavily on the Yugoslav economy. As Mr. Ruzin rightly points out, it is specifically the economic situation of

Macedonia that has the most impact on the lives of my young informants. Macedonia's industry has ground to a halt despite an abundance of workers, while there are not enough farmers working the land.

My informants entered university while there was a Yugoslavian state. They were to become Yugoslavian engineers, "intellectual" workers and an elite of the socialist state. Until 1994, the number of first-year engineers exceeded first-year enrolment in other faculties at the University of Skopje, but, in that year, more first-year students registered for economics. When my friends graduated, they were leaving a university in the Republic of Macedonia. The boundaries of their state had shrunk, and their freedom of movement and work opportunities had been greatly curtailed.

In the spring of 1991, the union of all engineering students of Yugoslavia called a meeting in Sarajevo to follow-up similar meetings held in different Republics every year. A week before my informants were to go to Sarajevo to meet old friends, the siege of that city began. Many of the friends from last year's meeting in Ljubljana whom they were looking forward to seeing, they never saw again. Some of them were killed or went missing. The security of the socialist state had fallen away; the jobs they had anticipated in the industry evaporated. Many had hoped to continue with postgraduate studies, possibly in other republics of Yugoslavia or abroad. Suddenly, this was no longer possible. Also, because the nature of their study demanded current knowledge in computer science and technology, the isolation of the Republic of Macedonia proved a heavy burden and many have now given up dreams of scientific enterprise. In their minds, they had nothing to contribute to the scientific community, and their state offered them no support in that direction.

ENTERING THE "WORKFORCE" AND THE LIBERAL MARKET PHILOSOPHY

My young informants knew they did not have the same security as their parents had had when they started their adult life; there was no longer a workforce or working community to enter. Their choices are limited not only because there is less work, but also because Macedonia is undergoing changes towards a free-market economy, and these changes greatly affect how many young people perceive the world around them. "So," questioned an informant in 1993, "the Western economy is about everybody fending for themselves and trying to gain as much as possible at the expense of others?" Even though people embraced the ideal of "Western," they were aware of the social consequences and distrusted efforts to turn Macedonia into a copy of Western Europe.

While in a café with friends, I saw three girls burst in, one girl in tears, all of them obviously upset. As often happens in Skopje, my friends knew these girls from the faculty and asked about their distress. It turned out that the girl in tears had just finished university and had started work with one of the private computer companies that were springing up very quickly. She had been very happy to get this job because the factories were closing down and private businesses or government ministries were doing most of the hiring. She had worked for three months for this company and banked considerable over-time. She had no signed contract, but her employers had promised to pay her after observing her work performance for three months. However, before the three months were over, she was laid off without any payment, and someone new was hired. This is not an isolated case. Another informant, a mechanical engineer, was offered a job installing alarm systems in cars. He was a specialist in Turbines, trained for a much more skilled occupation, but he needed money quickly because his father had just died and his brother had just married and there was another mouth to feed. He was offered 100 DM, around $40.[1] When he started working, his manager also employed another man to do the same job. The manager told them both that, after observing them for a month, he would hire the better one, who would start getting a salary from the third month onwards. My friend quit.

Few of the jobs my informants hold today give them sufficient income, and they remain largely dependent on the earnings of their parents. This group cannot achieve the important benefits of working, such as status and self-esteem. During socialist times, it was commonly believed that changes to political and economic structures would form a new society, a belief again prevalent among my informants. What they are searching for, however, are new values drawn from what they see on American television, values that they do not see in their own society. They believe that if these were the values of Macedonian society, success would depend only on an individual's own strength and knowledge: they could earn appropriate wages and recognition and afford the happy life they seek. They feel that, at present, this life can be attained only by emigrating to New Zealand, Canada, or Australia, or by holding on, interminably, for a better job.

SURVEY

Mr. Ruzin describes an expectation commonly held by those in post-socialist countries: a free-market economy will solve their economic problems and lead directly to a Western lifestyle. The disillusion felt after such expectations fall through is common. The standard of living in the Republic of Macedonia decreased under the free-market system, according to a 1994 survey by BRIMA (British Macedonian Social Surveys):

> [In] June 1994 32% of the surveyed people indicated that, generally speaking, they and their family's living standard had decreased significantly, 32% answered it had somewhat decreased and 30% said it remained the same; 1% stated that it had increased a lot and 2% did not know.

The free market was clearly not a system to covet. In the same report, "42% of the surveyed people agreed that the privatisation of state companies is the wrong step, 42% said it was the right thing to do and 16% did not know." Here are answers to the question "Do you think you will lose something if you don't participated [sic] in the process of privatisation?": "23% surely felt this, 15% thought it probable, 41% thought that they would lose nothing and 21% did not know." And the question "Do you think that the creation of a market economy which means the ending of state control is right or wrong for the future of our country?" elicited these responses: "32% thought it right in November 1993 as well as in June 1994; in November 1993, 51% thought it wrong as compared with 49% in June 1994, and 18 to 19% did not know."

POLITICS IN A "POST-COMMUNIST" SOCIETY

My informants certainly associated a better life with a "Western" style of living, and, in the summer of 1990, Macedonians thought both would result, believing that, with the reform program of Ante Markovich, "Yugoslavia will soon be able to join the EU." By 1995, many believed that a desirable future could only be achieved outside their own country. Immigration forms were filled out. People did not live with the hope of a better life, because Macedonia was regarded as a "post-communist" country and seen as lacking the quality of life that could make it comparable to other European countries. Macedonia was missing what Mr. Ruzin detailed: capital, a tradition or a history to relate to and be recognised under; and democratic and civil "knowledge" that would form a society in which to live. Sonija describes this perception in her account:

> ... This is the Balkans; you will never get any civilisation here. I have been to Australia, the Netherlands; these are civilisations, but not this. Germany has a history; you have kings, lords, writers, and musicians. We are peasants, we have been illiterate peasants for hundreds of years, a wild bunch of primitive people who lived scattered in the mountains. We don't even know where we came from or who we are: Slavs, even the gypsies are more interesting. And we are still peasants in our minds. You see it, men spitting on the street; or your hallway, it smells like urine; you know what I mean, all the broken mailboxes and demolished bells, the windows broken and everybody throwing garbage anywhere, basically out the window. This would not happen in Germany, would it? The Netherlands was so nice; everything was so clean.... And here we are and supposed to make politics. We don't know what democracy is, how could we? I was not happy in Belgrade; they were laughing at my brother in school. This is my home, but I would like to go to Australia. There are many Macedonians there.

People discussing politics were regarded as a nuisance, and if politics were discussed, the focus was not so much on history, economics, or democratic development but on the Albanian minority, "that they were getting too many rights, that the government was too afraid of them, and 'spoon feeding' them." Particularly after the shooting in Tetevo in February 1995,[2] feelings were intense about the issue of Albanians wanting a separate university: "The next thing they will demand is their own country." The Albanian insistence on being recognised as an autonomous group by Macedonia was a political issue that overshadowed everything else in those years, including the war in Bosnia and the economic crisis in Macedonia.

One evening I sat with friends in a crowded café called ZZ-Top. One of my friends had chosen not to come because she feared Albanian terrorist attacks; her friends were laughing: "Oh, she sees everything so dramatically. She even told us to tell you, you should make sure to leave the country. And she sells petrol to the Albanians!" Despite the many political issues at stake in "post-socialist" Macedonia—changing the political system into a democracy and reforming a marred economic system, changes that affected the everyday lives of my friends—people discussed the issue of the Albanian minority in Macedonia. They feared that the Albanian minority would want to split from Macedonia and join Albania and Kosovo, thus creating the mystical homeland of all Albanians called Illyira. In regards to some of Macedonia's own expansionist territorial claims, for example to the homeland of Alexander the Great, most people with whom I spoke chuckled at this vain idea or expressed concern that Macedonian nationalists could cause Macedonia to be isolated from Europe by pursuing such arguments. Nevertheless, my friends did not see that creating Illyria was precisely the same farfetched idea, and that, as I suspect, it was not an idea in the hearts of the majority of Albanians living in Macedonia at the time either.

Today, Europe and the UN are talking once again about ancient hatreds and historical land claims, forgetting that the real issue is the power vacuum created by the fall of the socialist world, a power vacuum that, we should not forget, Europe and the United States wish to fill. A friend explained in 1995:

> We have not developed a real social, economic, or political life. We are just moving all over the place. The main problem with the political parties, and it's probably the same everywhere, is power. The problem is that we are only three years away from separating from socialism. Many of the same people were running enterprises, still running the same things. Running our lives. I mean, it's normal. We're just floating nowhere, trying to grab what we can. We're always calling on the West. The West should recognise Macedonia. The West should help Macedonia, but I don't believe we are presenting ourselves well to the West.

Presenting Macedonia well to the West in order to gain access to the political and economic power that the West represents manifested itself, is expressed, I argue, in the obsession of creating a noticeable difference between modern, urban Slav Macedonians and backward Albanians. This difference was created by the images that Western academics, journalists, economists, and politicians alike presented of Eastern Europe. This difference was created in order to establish a new power balance, a balance that, once and for all, would prove the superiority of Western democracy. Consequently, Eastern Europe is looked upon and studied as a region of small-scale societies that developed into nationhood because of socialist agricultural transformations. Also, because critical studies of Eastern and Central European societies are few,[3] scholars stress economic desolation when looking at Eastern and Central European societies before and after communism. What is forgotten, however, is that, beside the socialist economic system's exacerbation of the region's economic deficiencies, Eastern and Central European countries did not receive anything like the Marshall Plan that brought Germany its *Wirtschaftswunder.* The result was economic despair followed by political instability, which the Western media describes as "backwardness" and the politicians as the inherent Balkan problem, the erstwhile "powder keg of Europe." We should remember, however, that today's conflicts and economic desolation are not only products of socialism but also products of the exclusion from power. It is the same exclusion that my friends fear today, readily endorsing the image of backwardness in order to distance themselves from this image (since they then talk and look at things as "Western Europeans").

Academics, for example, when discussing social change in former socialist societies stress its "uncertainties" or talk about a "bleak" economic outlook, perspectives that contribute to the image of East-European backwardness. Further, people like the Hungarian historian Ivan Berend have called Eastern Europe "the crisis zone" or talked about "an ethnic shatter zone." Michael Walzer, as well, has described Eastern Europe as the hotbed of "new tribalism." Then, with the outbreak of the civil war in Yugoslavia, the term "ethnic cleansing" became very popular with the media and added to the characterisation of backwardness.[4] This backwardness is reflected in the self-image of young people in Skopje:

> ... In my opinion, there is no legal contract, I mean, no legal system. The laws are not functioning. Maybe by starting with the laws, we will improve things. I don't mean by that that we need more police or restrictive laws, but bringing order, you know. Especially in business. To stop these, not quite crooks, but a lot of these companies working now they are ... corrupt.
>
> *Question: Why doesn't the legal system function?*
>
> ... I can't explain that. Many things are okay on paper but in reality they fall apart. For example, if somebody owes you money, as a company, to another company, there is no way for you to get your money back.
>
> ... There were some trials of people organising debt-collecting agencies. You pay them to beat this guy up or force him to pay, but that is not the legal way. It's not what we need. Maybe it works in the West also, but it's not the

> real thing. I don't know, maybe someone has an interest in that. Maybe we still need time for changes, maybe we cannot change the circumstances. There are a lot of reasons for how things are. You can't just blame the government or the people. It functions like that. If you see our surroundings: Serbia is at war, Greece is blockading us, Bulgaria is in a worse state than we are with much more crime or corruption, not to mention Albania. We are living in a ..., I don't know. (Informant 1995)

Macedonians' judgements of socialism's legacy are contrary to Pine's assessment that the socialist system in Poland was popularly judged by its omissions rather than by its achievements and that these judgements were formulated both in reference to a historical past and a contemporary contrast to the West.[5] In Macedonia, the socialist system was respected for its social achievements, and elements that were seen to be reminiscent of Macedonia's pre-socialist, peasant past were contrasted to the West. Those elements, seen as "Balkan," existed before and during socialism and exist yet today: only the reference for these "Balkan" elements has changed. The connotation of "Balkan" has changed over the years. Initially, "Balkan" suggested a peasant past and oppression, then "Balkan" became the antithesis of the ideal socialist, industrial state, and today it is contrasted to what is seen as the West. I would argue that, as a political consequence, the Albanian minority came to stand for this "Balkan" element within Macedonia. And finally, the political hope for Macedonia was vested in becoming closer to the West.

INSECURITY AND UNEMPLOYMENT

It was common for an engineer still at university to become employed full-time. A friend of mine had to leave university to work because her father, despite having the same training as she did, had been laid off, her mother was a lower income earner, and her father's drinking was expensive. Her most employable skill, however, was her facility with English and not her engineering abilities. Others left before they graduated for different reasons: family issues and personal beliefs—"... war will come, so why should I study?" (Informant 1992). As engineers, they were not filling a need in their own society, which put all its economic effort into trading, mostly with Turkey or Germany and illegally with Serbia. The factories that once employed engineers were closed. Often fathers were laid off, and mothers would finance their families alone, for example, on the income of an economist, a common career choice for women in the parents' generation. The mother, then, and sometimes the daughter with her language ability[6] were able to work for one of the foreign agencies flooding into Macedonia, and they became the only income earners in the household, in addition to their responsibilities of running that household.

> Yesterday my mother and I did the weekend cleaning; my sister had to study. I asked my father to help me to put up the curtain as I could not reach, and he screamed at me and told me that this was my job to do and not his. He sits there all day and does nothing. Watching television all day. He would not touch a thing in the household. My mother is working overtime at the company to get some extra money, comes home and then has to do the cooking, cleaning, washing and ironing. I cannot understand how she can take this. And then she wants me to marry! (Informant 1994)

These changes in the economic and social lives of my informants are directly linked to economic and political transitions within the Republic of Macedonia. In the past few years, the Republic of Macedonia experienced an increase in marketing and sales and a decrease in production. This decrease in production is related to the closure of factories that produced parts for assembly elsewhere in Yugoslavia. Since the Yugoslav union does not exist anymore, and Macedonia does not have the capacity or the capital to build an independent infrastructure, the factories are closed and the unemployment rate has risen to 30 per cent. Funds offered by EU and US government-sponsored programs are used to subsidise large programs of government spending. Often, this money is spent in ways that are far from useful, and the work created is neither demanding nor rewarding. In government-sponsored projects, the young engineers work at the level of technicians, and their intellectual potential and acquired skills are not well utilised. For instance, I was told that one ministry received a 10-million-DM subsidy from a foreign aid program and used this money to buy a computer system. The deal was made between the ministry and a private computer company belonging to the brother of a very high official in that ministry. This computer system was the newest on the market and offered opportunities of which some of my engineer friends could only dream. However, this computer system was never put to use as none of the responsible staff had the expertise to run it.

Their potential unfulfilled and society making limited use of them, my informants increasingly demand leisure goods and services that promise them some reward. They want something to bring them closer to their dream of a new society, one closely resembling those they might have seen in Western European countries or on American television. A good life is equated with buying things that seem to represent images of happiness.

LIVING ARRANGEMENTS

Nearly all the young women in Skopje live with their parents, with virtually no chance of having their own place, even if the family owns another property.[7] This situation arose during Macedonia's transition from a socialist to a free-market state. After 1945, in socialist Yugoslavia, most young people left their homes in the villages or small towns to move to Skopje to work in state organisations and to move into their own flats supplied by the state. But today young women live with family, and a lot of tension arises from this situation and their desire to emulate the life of their West-European counterparts:

> They treat me like when I was sixteen, they ask me all the time what I am doing and where I am going. I do not feel like telling them what is going on at work, but they want to know everything. And my mother is really worried because I mentioned Stefan so many times, and he is married. I hate it when they control me like this. I think this will only change the day I am married, and it may not even change then. You are lucky, you can do whatever you want and nobody tells you what is right and wrong. You are your own person. I would like to find out things for myself. I would like to make my own mistakes.

Another friend describes her situation:

> It is not a feeling of being controlled; I cannot describe that feeling accurately. I do things because I want to do them. They complain that I haven't done things correctly or my mother says: "Oh, you came in so late last night," and they say, "that is the reason why you feel like this." When they know things, when they know why and how I should be, and I am not that way, then it is not a feeling of being controlled, it is not … I don't know; I would do the things anyway, like that. I see unnecessary quarrels with my parents, and this just puts more tension in our house. My mother starts screaming, I start screaming, my father starts after a while as well, trying to calm me down. If he is trying to calm me down, he makes me even more nervous; it is just unnecessary. I do not like the way we get into fighting, and that's how it is for a while at home now. It is too much, staying up late, going to work—I have obligations at home—and yes going out too much, staying out too late, yes. I am not getting enough sleep, and then you are not really ready to do what you should do.

Parents see their daughters spending most of their money on things they feel are unnecessary, and they see them coming home very late. They were concerned about their daughters' lack of sleep, and, most of all, they had conflicts because they wanted their daughters to find a nice boyfriend but not to get "in trouble"[8] and have people "talk." They knew they were supposed to be "Western" parents and tried their best; their daughters, however, felt strongly that their private lives were not their parents' business but their own responsibility. To the daughters, this thinking identified them as "Western," modern, and individualists. They insisted on the right to make their own mistakes. Nevertheless, these young women seemed to demand autonomy only on their terms. Few of them took this autonomy further into more uncomfortable areas of life that demanded effort on their part: moving out, finding a job by themselves, or emigrating. They relied on their family for support, and it was at this point that the family exerted influence over their decisions. A more complete independence from the family could result in social exclusion, as the example of a young female informant shows. She asserted her independence and, taking her two children, left her husband. While her sisters supported her, the rest of the family turned away from her.

MONEY

Economically, my friends were not able to fend for themselves and were dependent on their families. Even though they had their own salaries and were not expected to relinquish them to the household, they nevertheless did not have complete control over their own finances. They were expected to supply their siblings with pocket money and "extras," for example jeans, sweaters, or shoes. Providing these things, far from being a duty, was regarded as part of "sibling" affection. Nevertheless, parents exerted subtle influence on their children's spending through little comments about wasting money: "You go out every night. Are we rich? And the new sweater, and the new shoes!" The daughter's perspective was somewhat different:

> My mother is always telling me I should not go out because it is so expensive, and I should stay at home and learn to cook and help her in the flat. She goes on and on about how lazy I am and that I am not doing anything useful. Yes, I am lazy, but when she is getting so upset saying that all the work is hers and everybody is just going out and leaving all the work to her, this only makes me want to go out more. I know this is not fair. I just cannot stand her nagging. She wants me to stay at home, and then she complains that I have no boyfriend and asks me why I don't have one. She says that I should stay home and learn how to cook. Well, either I go out and look for a boyfriend or I learn how to cook, but somehow this does not make sense. And she goes on and on about the money. She wants to exchange all my money into *marks*, and then she hides it.

Some time later, this same friend decided to spend most of her salary at one of the beauty salons. It was very expensive, and she told nobody the exact cost. Friends of hers who frequented the same beauty salons also concealed the real prices of these places and how much money they were spending. Nevertheless, Stefanija's mother one day went through her wallet to count her money and started to scream about where Stefanija's salary had gone. The ensuing argument was about "household responsibility." Stefanija came to my place very upset:

> I know it is not right to spend all this money and hide from everybody how much I spent. I was telling my mother that it is not her business what I do with my salary. But my mother went on that it is very much her business as she is saving money for me, buying *Deutschmarks* for me! Then my father tried to calm us down, saying that he thinks, for sure, I will tell them where

> the money is, later. Really, I never would have thought that my father would care about these things, that he actually thinks as well that it is my parents' business where the money goes, but maybe he was only trying to calm my mother. When they will ask me later, I will just tell them that I loaned the money to someone.

Then she borrowed $25 from me to buy a pair of underpants for her boyfriend.[9] She went on:

> I know I am not right. I know my mother is right in saying I am still dependent on them and that in some ways it is their business what I do with my money. But I think I have the right to be treated like an adult. I told my mother this, and she told me she would only see me as grown up when I am a married woman. Wonderful. Of course my mother at my age had long since been married and had a child of four; nobody told her anything. She never had problems like I have. It is really sad that I fight with my mother like this; it can be so nice with her sometimes, but we make each other more and more nervous.

Such dependency is not unusual in times of economic crisis. The parents of my friends had it, in some senses, easier than their own children did. After 1945 and after the earthquake in Skopje in 1963, the city saw a great influx of young people from small villages and towns. These young people came to build up the fledgling socialist state. They took on responsibilities that their children today would never consider. Stefanija's father left his small village in Eastern Macedonia when he was 16 years old and came to the city to be a teacher. He got married, his young wife moved with him into their own inexpensive flat provided by the state, and she started to work as an economist and look after the household. Their parents were far away and too inexperienced with urban life and with socialist society in general to provide advice. The young people who came to Skopje were employed by the socialist state and did not need to seek employment through connections, not that their parents could have provided such connections anyway.

If Yugoslavia had continued to exist, what would my friends' situation have been? It might very well have been the same as it is now, but it is important to note that they had not expected their current circumstances. In their thinking, their lives should have been like those of any middle-class German or English student. For their postgraduate studies, they might have gone to a different university in Belgrade, Ljubljana, or Zagreb or perhaps even abroad, and gained further independence from their parents. They might have been able to get a cheap flat, and they felt that certainly they would have had their "own place" when they got married. Marriage or postgraduate studies would have been their way to greater independence, an independence they felt their parents had been given but they were refused because of a situation not of their own choosing. In comparison to many other post-communist societies, Macedonia came to independence not as a matter of real choice, such as the choice people perhaps felt they had made in Croatia, Slovenia, or Serbia. Rather, Macedonians had independence thrust upon them when Yugoslavia fell apart. Their only real alternative was to be swallowed up by Serbia, which had once been an occupying force in Macedonia.

CONNECTIONS

In Skopje, no one I know got a job without connections although sometimes these connections were concealed. Some of my friends tried to convince me that it was solely their credentials that got them employment. However, even with their credentials, they still needed connections to beat other candidates who might not have had the credentials but who had the connections. The nature of the job determined the extent to which connections were a factor. It was easier to get into private business with credentials because sometimes expertise, in particular facility with foreign languages or being a specialist on a specific computer program, could outweigh the "connection factor." According to BRIMA's 1994 survey—and my own knowledge, which can confirm this—to the question "If you want to have success, you must have 'connections,' in our country?" 90 per cent of respondents answered "agree," 4 per cent answered with "Don't know," and only 6 per cent did not agree.

I asked a friend who was running a music store some questions about connections:

> *Question: How important do you think connections are for being able to have a business?*
>
> ... Here? Oh, I could talk a lot ...
>
> *Question: Is it possible to have a business without connections?*
>
> ... Probably not. Many of these rich people around who have big companies now use their connections from the socialist times. Many of them were managers in state enterprises and they just transferred the capital with their connections to their private companies. Connections are very important. I have many friends, lots of connections, but I never use them for my job because it's a special kind of job, you know. I have friends everywhere, but people cannot really help me with my business. As I told you, we are probably the only people who deal with this material in a legal way.[10] So, it is important to have connections, especially here in Macedonia. Probably everywhere, but here in these circumstances nothing would even work without them.

In addition to getting a job through connections, knowing someone ensured that one would not be laid off easily, even out of economic necessity. A strange effect was that some engineers became teachers or economists because these were the connections their parents had; the serious issue of relevant expertise for these jobs was irrelevant:

> They do not care what you are, the only thing that counts is "who you know." In my mother's ministry, they employed a woman who was in her eighth month of pregnancy. Nobody has ever seen her since she is now on maternity leave—and she is getting 80 per cent of the salary from a job that she has never done! And you ask if women are discriminated against! What kind of job I get depends solely on the people my parents know.

In summary, these young engineering students leaving university and entering a different phase of their lives were far from becoming the self-determined agents they had once dreamt they would be:

> I want to be independent, to have a job that is challenging, colleagues who are fun. I do not need a boyfriend. I will have my own little place, or Ane and I will move into her flat together. It will be so nice to do whatever you want. I would like to have my own room, decorate it, as I want to, have my own stuff. On the weekends I will go for lunch to visit my parents or maybe invite them over. This would be nice.

Instead, they were thrown into greater dependence than they had ever expected from their perceptions within Yugoslavia or their ideas about Western Europe. They even saw their parents as having more freedom than they themselves had. Although *vrski* (connections) were important in Macedonian social life before and after the fall of Yugoslavia, today *vrski* cannot guarantee security, money, or living space, the lack of which has become the hindrance to what young Macedonians aimed to achieve. This is what my friends faced when they started work:

> My father had to change his job because his company was completely run down. He and his colleagues went to work just to keep their jobs, although there was no work for them to do. They did not get paid anything. He is now working for an electric company. He has a very low position, like I have. He had no other choice: he had no social security, no health security. Such things theoretically existed, but his factory did not have the money to pay for that. The state does not pay either. He needs his social security payment for his pension, which is another reason why he changed jobs. People are very scared to lose their jobs. We hope it will become better, but they said that this year it would be very, very bad. There will be an enormous number of people who will lose their jobs. Everyone is afraid that they could lose their jobs. The government is supposed to pay unemployment insurance, but the government does not have enough money. The money which is in the country, is not regular, not for taxes.[11] The money one can see is not money, which is put somewhere, in an account or something; there is only cash in hand. The taxes are up to eighty percent. That's why the private

> companies do not pay any taxes. And the government is not strong enough to enforce such laws. It is very easy for the private companies; they pay 3000 or 4000 DM to some official in the government instead of paying 10000 DM tax. It is simply the way things go here. There is an organised Mafia in the government. And people are afraid, they are afraid that they will lose their jobs. That's why nobody does anything about it. My father is the same. He only talks and talks. I do not like this about people here; they only talk and never do anything. People are afraid. We would not have been 500 years under the Turkish Empire if people had been different. (Informant 1995)

This is the world that Nela, Ane, and Beti stepped into when they graduated in 1993 and in early 1994. At this time, they were around 24 years old, had been born in Skopje, and had lived there all their lives. Their parents had come from other towns and villages in Macedonia. When they started working, they had no boyfriends, and the support of their friendship group was great, although it weakened later. Some of their friends were still finishing their last exams or graduate work or were looking for jobs. Besides the student dilemma of having to balance time for sleep with going out to the cafés in the evenings, they felt uncertain about what they wanted. Nela and Beti were sure at that time that they wanted to continue their studies with postgraduate work. They saw their work as only a filler, helping them to gain experience and some money for future studies as well as to find an appropriate dissertation topic. At this time, Nela and Beti were still paying graduate fees to the university and saw themselves as students and not different from the full-time students at university. This status implied going out a lot and taking part in the "student games," similar to student Olympics. Ane took her time finishing her graduate work and then refused to have her father organise a job for her at the television station where he worked because there were only "old people there."

THE CASE OF NELA

Nela graduated in the summer of 1993 and then applied for a few jobs that were advertised in the newspaper in a *konkurs*, a competition. I was told that her application did not make sense as these jobs were only in the newspaper because the law required that they be publicly advertised; the positions had already been given to "friends." Nela applied for a job in customs where the pay was supposed to be very high, in order to reduce the prevalence of bribery. There was another job at the post office at that time, but as she told me later, an Albanian got the job due to the policy of "positive discrimination,"[12] even though, according to Nela, the Albanian's credentials were not as good as hers. As her job applications failed to result in employment, her father took action. He organised for Nela a job teaching computer science at three different high schools: agriculture, art, and medical schools. The school of agriculture lies on the outskirts of Skopje and draws those students who could not qualify for any other high school, so they are educated to become farmers. In the school system Macedonia inherited from Yugoslavia, high schools specialise in future professions: there are technical (engineering), language, medical, art, and agricultural schools. The brightest students, or those who had bribed their way in, were supposed to be found in medical schools.

Nela enjoyed her work in the agricultural school particularly because her colleagues were young, whereas in the medical school staff were older. Older people held the more respectable and rewarding jobs; the less rewarding ones were given to younger people. In Macedonia, a post-communist republic, my friends commented on this situation many times, as they felt more competent than those from the "old communist times." They felt that those older people had been hired only because of their party membership and not because of their abilities, especially in the field of computer science. My friends rather liked working with young people and did not feel they could learn from older colleagues, whom they feared were interested not really in work but only in enhancing their own power and welfare. My friends often said these people avoided work and commented that this was a "Balkan attitude."

In contrast to such "Balkan" attitudes were those of a woman of forty whom Nela admired. Although the wife of a priest, this woman had many young lovers and was enjoying her life, living it the "Western" way. Even though she was considered vulgar, for Nela she was honest and full of life. Nela contrasted this woman and her younger colleagues in the agricultural school with her older colleagues in the medical school who were accepting bribes from their students for exams, following the "Balkan way."

Nela also felt that there was another difference between these two schools. She felt that each school reflected a different work culture, one she considered Balkan, the other Western. Whereas in the agricultural school, business was more relaxed, in the medical school there was an emphasis on performance. In respect to Nela's salary, these differences became apparent. The agricultural school still owes Nela one month's salary, though she has been there to see the director many times. Her father, a "friend" of the director, had even phoned a few times to see about his daughter collecting her pay. The school was always ready to give her the money, but every time she rode the bus forty minutes to collect it, one of the people required to sign for or arrange for her to receive the money was "not in."

In contrast, at the medical school she was disciplined for not reporting marks for her students on time, and her salary was reduced by 10 per cent, which occurred for most of her colleagues. In both places, she found the work very boring, as she had to teach students the basics of communications and computer science without any computers at hand. Although Nela felt that the agriculture school had a "Balkan" work attitude, which she herself detested, she felt more comfortable in the atmosphere of this school and with her younger colleagues, which she felt were more "Western." In the medical school, she enjoyed the work attitude, which was directed towards performance, but did not approve of the "Balkan way" of accepting bribes. The attributes of "Balkan" and "Western" overlap in these schools and illustrate the contradiction many of my friends felt at that time: whereas the younger, more "Western"-oriented teachers were working in an old socialist working environment, the "Western" working environment was occupied by the older socialist elite.

Though Nela liked her students, she felt that teaching was not a challenge. She did not take her work very seriously, and we helped her mark the essays. What she liked most about her job was the free time she had, though she complained about having to get up so early. However, she still believed that she would continue her postgraduate studies because she knew that another degree would be helpful when the time came for her to leave the country.[13]

This difference in the two schools is significant as it demonstrates the dilemma my young informants felt. They felt that they had been the elite of Yugoslavia as the young generation that had finally overcome the stigma of "the Balkan" and was worthy of living in "the West." Given the disintegration of Yugoslavia and the new distance of the West, significant changes occurred in the Republic of Macedonia, especially in the economic sector. Foreign aid was freely handed out to any sort of enterprise, to schools and universities and to the government. Many foreigners had come to Macedonia and were living and working there. However, if Macedonian graduates did not work in one of the private agencies with the foreign companies and opted for the more secure jobs in government, they were faced with socialist elite mismanagement, as they termed it, and the squandering of the possibilities given to Macedonia from the "West."

Although Nela finally got a job in a government ministry, it was not always clear that her father had strong enough connections to achieve this result, and he had to contact a number of people to ascertain that his connections were, in fact, stronger than those of competitors. They had a "friend" in the ministry who let them know that Nela was suggested, but everyone was still unsure about whether the other woman who was short-listed would get the job. The other candidate "knew" another very important person in the ministry. So Nela's father made some further contacts, and Nela finally got the official offer. Two months after she had started, she was already complaining that her bosses would not give her any interesting work. She thought this was because her department head was disappointed that his "friend" had not received the job. Even though she began to get work to do, she found it boring as well and began looking for another job. This decision was highly criticised by her parents and other family members, who told her she could not switch jobs like shoes. Nela decided to immigrate to New Zealand and was busy making the necessary arrangements. She gave up the idea of continuing her studies in Macedonia and decided to do her postgraduate studies in New Zealand after she had earned some money there. She was still periodically checking the newspaper for *konkurs*, job advertisements, but was disillusioned as they were "fixed" already. To actually get a new job, she would have to draw on her father's connections again. She felt that, outside Macedonia, life would be very different in many positive ways, but also realised how difficult it was to obtain a visa that would allow her to leave Macedonia. She has resigned herself to her job and claims to be used to the situation of not actually working. Nela has learned to use the time at work for private endeavours as everybody else does. She claims not to feel guilty because of this any more as she explains:

> I don't know. Things are somehow slow here; not only the ministry, this is the case for others as well, but that is the way it works here. I still would like to find a different more fulfilling job, but it is difficult. Some people here do what they feel like doing; some people work.

To find pleasure in what one does is seen by all of my young informants as a "Western" attitude. Unrewarding labour or a way of working that is distinguished by its inefficiency is seen as "Balkan."

THE CASE OF BETI

Beti took the initiative of finding employment for herself. She had worked abroad in Germany and England a few times over the summers and had a clear idea of what it meant for her to work. She applied for jobs in foreign embassies and relied more on her language abilities than on her engineering experience. She was unsuccessful with the embassies but managed to find work in a private computer company where she was to design programs. Although her salary was not high, she was satisfied because she saw her work as experience that would get her further in her studies. She subsequently applied for a job at the post office, but then decided that it would not be challenging enough. She then applied at the defence ministry and got a job as a computer technician, as she said because of her knowledge of German and English and some good "connections." She also told me that she was sometimes not treated well because some "friend" of her boss had not received her job. They owed her money for the trial period of two months, but she had to accept no payment for these two months in order to avoid stirring up trouble. She told me that she felt that companies now do not like to employ women and that she had been very lucky.

Beti has no female colleagues. Her immediate colleagues are nice and friendly, but she has some difficulties at the ministry because she is not married and some men make unwanted romantic advances. A friend told her that the men who are acting overtly are only very outspoken and that, basically, all her male colleagues felt similarly about her. For a long time this disturbed her enormously, although she took pride in being the only woman working there and was flattered by the attention she received. She dresses up for work, though she emphasises that she does not act like the "pretty women" because she is an engineer, and she walks around in jeans. Nevertheless she likes her work and the aura of importance that working for "internal security," with its special rights of the police force, gives her. The situation is not stable in her ministry though, and a few times there have been salary cuts. Consequently, she is looking for another job without her parents' knowledge. Beti feels that her parents only want her to be employed at anything and then get married to fulfil her proper destiny. Beti was short-listed for a Swiss company that wanted to open an office in Skopje even though they were mainly interested in people from banking businesses. Beti was interested because the job offered a much better salary than an engineer could get. Again it was her knowledge of languages that allowed her to be considered for better work with a foreign organisation. For a long time, she felt it would be to her advantage to take a job with the Swiss company if she were offered one. However, she decided against it because she felt more secure in the ministry where she already had a job and where the chances of her being laid off were not high. She noted that unemployment would be disastrous and that she would not know what to live on if she became unemployed. She was in a secure situation, though, because her mother, at a great sacrifice years ago, had bought, against the will of her father, three flats. Her mother, who worked at a bank, had insisted on this investment for her two daughters. The last flat was bought just a few days before all bank accounts in Macedonia were frozen. At that time, her father was unemployed.

There was virtually no unemployment insurance, and people had to rely on their own family resources. Times could be very hard for some families, as they were for one of my friends whose parents both became unemployed and were renting out their flat in Skopje to UNPROFOR. They moved in with my friend's grandmother who was living in a small town, and my friend had to quit university because she had no place to stay in Skopje. All five of them were living on the meagre pension of her grandmother and some savings. Talk of this particular situation made Beti reconsider her decision to work for the Swiss company; she feared that a foreign company could pull out of Macedonia at any time. By changing her mind, however, she felt she was again giving up a chance that she might, one day, be employed in a Western country.

Her decision did not take into consideration that her life-situation would not have changed dramatically if she had become unemployed. If, however, her mother, were to be laid off, the consequences would have been dire for Beti's family because her mother earned enough money to support the entire family. To acknowledge this situation required Beti to recognise her own dependence, which she did not easily accept. She felt superior to her parents who had never been outside Macedonia whereas she had worked and travelled abroad and knew what "real" life was about. Still, she was not satisfied with her salary, despite the general difficulty of earning much in Macedonia by honest means, as people kept saying. She felt she was not valued highly enough in her work as an engineer and believed that technicians, the socialist "working class," were still valued more highly. When asked about her university education, Beti said she still wanted to continue her post-graduate work, although doing so would bring her no financial advantage in the long run in terms of a better job or promotion. Besides being eager to learn, she secretly hoped that further university education would increase her chances of working abroad, as well as give her more work experience. She complained many times that education was not valued highly enough. She pointed out that there was a law designed to give her department head, who had only a high school education, the same salary as a beginner with university degree. But she doubted that such a law would ever be enforced. For her, too many people with little education were sitting at the top:

> That's what is left from the old times. They were good party members, and that's how they got their positions and the same for their children. They use their positions to get high positions for their children. They have good connections, and they do not do anything; they barely come to work, and they get the same money as I do. Who has the most money? It is the people with low education. They use illegal ways to get so much money. (1995)

She constantly dreams of leaving Macedonia because she feels that her country, in its present transitional situation, cannot give her the support she feels she needs:

> The young generation is the lost generation. They have the knowledge the new times need, but they are missing guidance. The older generation is only busy securing high positions; they are not interested in work, and they cannot be fired— they determine the rules of the game. With them in power, we never will manage to change our country.

At the moment, however, she has a secure job and earns an average monthly income of 300 DM. Her parents are renting out two flats. Her father has a new job, though he has the same position and salary as his daughter after being employed for over 30 years. Economically, her family is better off than before Yugoslavia's disintegration because her mother had planned with foresight. Beti's biggest concerns are the arguments she has with her parents who want her to be married now. Beti does not want to marry a Macedonian man because of the "Balkan" attitude of these men toward women. Being a "Balkan" man, for my informants, means to be chauvinistic and to have women do all the work. It also means not showing emotions and going out with many women, driving fast cars, and drinking a lot.

These men are seen not to care for women, or if they do, only in a sexual sense; they do not regard them as friends or partners. Men, however, accuse the women of caring only about money and fast cars.

However, because Beti hopes to leave Macedonia, she feels she must not be alone, as immigration officials in Canada, New Zealand, and Australia prefer couples and because she would be scared to be alone. She is applying for immigration to Canada now since another friend of hers has just had his visa granted.

THE CASE OF ANE

One Saturday, Nela and I went to visit Ane, another engineer, at the television station. She was sitting at her desk enjoying the warm April sun streaming through the windows. It was spring; people were moving around, coming in and out of the building. Lela, who has her own office because she is the daughter of a good friend of the father of Ane's boss, had visitors as well. I went to Lela's office to look at holiday pictures, and Lela was complaining that she was bored. In the winter, she had still wanted to find work as an engineer, but now she had become a businesswoman. She had settled for this, in response to pressure from her father. She could come and go whenever she wanted and complained that, even though she was hardly seen in the company, nobody cared, and she felt that nobody needed her. All of us had a nice chat and then went to sit with Ane on the couch in the entrance hall. We were drinking Coca-Cola and joking with the porter across the hall. We talked until Ane could go home and, although she left slightly early, the work she had to have done for the day was long finished. She was arranging television programming and had to fit the programs and advertisements together and time everything. Sometimes we would work together, subtitling a movie for some extra money. We strolled home and went to the pizza place to have a huge ice cream. In the evening, we all met again in the cafés. It was like a normal weekday, except that it was a Saturday and spring; there was a weekend ahead.

What made Ane different from her friends was that she enjoyed her job and did not "live for the night" as some of her friends did. Her work demanded neither special skills nor an engineering education. In reality, anyone in the company could have done her job, as she told it. However, she had discovered that she was a very talented organiser. Everybody in the station could, and did, rely on her. She used her spare time to keep a supervisory eye on everyone else. With her education, she had a special position in the station, and her immediate supervisor would often consult her about serious work. She started to read about the business, learned a lot about music, and took advantage of the free tickets and entrance admissions to the various cultural events a television station had to offer. She felt she had missed something in her education as an engineer, a view she liked to underline to her friends, who took it as an insult since they interpreted Ane as saying engineers were *ne kulturni,* uncultured. Ane enjoyed the freedom the station offered but took her tasks extremely seriously. Most of all, even if there was not much to do, she enjoyed working with her colleagues, all of whom were young and enthusiastic.

When Ane started to look for a job, her father, who had not managed to get her employed at the television station where he worked, organised a voluntary job at a private television station for Ane. At first, she did not want to work there, because the boss of the station, who was her age, was known to have a very violent reputation. Her father did not show any understanding of Ane's feelings and thought she should be happy to have a job. At the time, Ane did not feel she had a choice and decided to gain experience at this television station to prepare her for "real" work at the national television station. She spent considerable time in the cutting room and was eager to learn about technical support. Currently, she is the only woman in the division, and the supervisor of this division is her father's friend. She points out proudly that it is she, a woman, who is doing the "real" work to keep the television station successful, whereas her male colleagues are cleaning the machines. Although Ane is popular, some people are jealous because she has a diploma and is so young. Ane, though, despite her diploma, has no job security and has never signed any work agreement, does not earn much, and, most of all, does not have a job that qualifies her to be registered for a pension. She works voluntarily at the whim of her boss, who can decide to pay her or not or fire her whenever he wants. Consequently, Ane is still looking for a different job and hopes that, by performing well, she will be promoted to a contract position and get more demanding work, suiting her education and knowledge.

In the meantime, Ane applied for work at the post office at the urging of her parents. Her parents did not have the best connections, but a good friend, who was *kum,* godfather, or, in this case, best man at her parents' wedding, is a very rich business man who has made considerable money in Switzerland. Her parents hoped that he would have enough influence to help Ane get the job at the post office. Ane was not very happy about this situation because she liked her current work, her new interests, and her new friends. Most of all, Ane feared that the job at the post office would be boring, which would be harder to bear than the insecurity of her present position. Ane felt working at the post office would be so boring because there were many older people working there. The same argument surfaced when her father suggested that he might arrange some work at his own company. In May, she had an awful fight with her mother who could not understand why Ane refused her father's offer to help her find a job at his national television station or why she wanted to stay in the current insecurity of the *mladi raboti,* youth work, working scheme. In tears, Ane complained that she and her friends couldn't make their own decisions:

> I am 25, and my parents determine my life as they want. I have no say. I told them that I like the job and that I am good at it and I would like to wait until they give me a better job opportunity—a real job. In my father's company, there are only older people, and at my station everybody is young, and I love the atmosphere. Aneta [a colleague her age she befriended] is there as well, though this argument I could not give my mother. My father is upset because he is retiring soon, and then he won't be able to exert much influence on my getting a job there. But I don't want to have security now. I am young, so many things can happen. Why do I have to arrange everything for the rest of my life now; why can I not wait until life comes to me and offers me something? My parents only feel that I am acting irresponsibly, that I am not taking the responsibility they expect from me: I do not have a boyfriend, and I am not looking for one. They also blame themselves that they could not offer my sister and me a better life. This is why they try to arrange our life so it would be safe; they try to save us from their worst fears: insecurity and instability. I want to be independent, but I cannot be. I would like to make my own decisions, but my parents insist that they know what is the best and that it is a luxury to make one's own mistakes.

Ane was unhappy about her parents' decisions but did not dare insist on her desires. She spent many hours thinking what she would do if she really did get a job offer from the post office. She knows that one day she will have to accept something more than voluntary work, but for now she does not feel threatened by the insecurity. Friends a bit older than she is do leave the private television station to work in more established companies, but Ane feels that, for the moment, she is doing the right thing. She enjoys her work, and she learns new things; she does not want to think about tomorrow. Today she enjoys what she can experience in the present. Security is not an issue for her because she feels that there is no security anywhere and that one depends only on the goodwill of others. Her dream would be to become the station's technical director and, as she says, there is plenty of time to achieve this. Ultimately, Ane's ability to help everyone and her likeable personality are the only assurances that she will keep her job, be paid, and get holidays, all of which are dependent on connections and not on a work contract. The boss's father puts restrictions on the budget and does not allow Ane to be employed regularly. She was scared to tell her father that her promotion was rejected on these grounds. He listened calmly and finally took the position that Ane should learn as much as she can while watching out for a better job. In May, Ane hoped that she would find work in other private television stations that were to open soon. She hoped that the friends at work would know people and could organise a job for her or would even transfer with her. Another television station, the International News Organisation for Eastern Europe, which transmitted its programs across Europe via satellite, was considering opening an agency in Macedonia. Only ten people would be employed, and the salary was expected to be very high. Ane was scared to talk about this job, as it would fulfil her wildest dreams. With such a job, she might even have the opportunity to work outside of Macedonia.

CONCLUSION

I have presented only three cases at some length, but all of these young informants are in very similar situations. They eagerly anticipate change but feel they are being held back by the transitional state of their country. They oscillate between optimism and pessimism. Social change may generate great anxiety because the future is suddenly much less secure, but there is also the possibility of experiencing this change as gaining "the meaning of life" and not only as loss (Marris 1974: 148). On the one hand, my friends hope for a better future and are looking forward to being given demanding tasks that would allow them to further their ambitions while, at the same time, they are reluctant to believe that today's world will change dramatically. They still see the same old attitudes towards work, which never met the official socialist picture of the working class, and corruption is still prevalent. Their reaction is to hope for emigration to a classic capitalist country where their dreams of freedom, independence, and consumption promise to be satisfied. Today, people in Macedonia do not face the supposedly typical capitalist situation of "no pain, no gain," but experience instead "pain without gain." In addition, the Republic finds itself in an ideological void.

Writing about "getting along" touches on many aspects of my informants' lives: how they "get along" with their parents, through adolescence into separation; with the world around them; and with each other in a time of dramatic change and terror because of the vast unknowns in their lives. There are losses, but also hopes and dreams. Macedonia has avoided direct involvement with the Yugoslavian conflict and, despite internal ethnic tension, has withstood the anticipated civil war so far. A new society, one that people want to believe in, has been formed. The memory of the socialist past is honoured through monuments and mementoes of Tito, and there are no attempts to rewrite socialist history. The Republic has experienced transitions from "feudalism" to "socialism" to something not yet defined. It is not correct to speak of post-socialism, but neither is the country socialist. Macedonia is evolving, and its definition lies in the future.

My informants entered a new phase in their lives, finishing university and starting employment in a time that coincided with the political change from Yugoslavia to the Republic of Macedonia. In comparing the "coming of age" of my informants to their parents' adolescence, one comes to the understanding that both groups see themselves as "the elite." The parents played an important part in Macedonian society after 1945 through hard labour and by accepting the responsibility of transforming a peasant society into an industrial nation. Their children, during the last years of Yugoslavia, saw themselves as an elite because of their advanced knowledge and their "Western" attitudes, which bridged the gap, in their eyes, between Western Europe and the Balkans. As their parents changed society through socialist ideology, my friends intended to change contemporary Macedonia through their unique qualifications. They saw their parents' generation as still bound to Balkan principles through Yugoslav socialist ideology. In the same way, their parents wanted to better the lives of the previous generation, which they viewed as composed of downtrodden and oppressed peasants who had to be brought into modern, industrial Yugoslavia. My friends wanted to lead their society away from this notion of the Balkan, the backward, non-European, *ne kulturni*.[14] This is a typical generational conflict: the younger generation wants to improve the world of the older generation, and both generations have a specific ideology about employment. Whereas the parents' generation saw employment as labour and received power through access to employment that could change society by party membership, my group of graduates viewed employment as a career in which one advanced through several jobs, gaining power in the process, and improved society by improving oneself. Both generations, that of post-war Yugoslavia and that of "pre-mortem" Yugoslavia, were generations of change. In post-war Yugoslavia, power to change had come through ideology, in "pre-mortem" Yugoslavia, through knowledge. In the parents' generation, security was granted by the state, but the "pre-mortem" graduates expected their lives to be secure because of their education, experiences, and qualifications.

Today this situation has changed drastically. Young people do not have the same influence on society because external politics is out of their hands. Today neither socialist ideology nor qualifications hold the key to success in Macedonia, but connections do, an element that all my informants identified as "non-Western," as "Balkan." Today, people see employment as having a "job" that holds no power. My young informants see the older generation misusing knowledge and assistance from "the West," again pointing towards the categorisation of Macedonia as "Balkan." In the following chapters I will examine the many ways this specific group of professional women graduates are understanding and altering their lives and thereby "taking charge" of their society and their lives.

NOTES

1. To give a rough idea of how much that would be: £10 to £15 would pay the electricity bill for a month for a four-person family. An average wage for a fellow engineer working in a ministry would be three times as much.
2. In February 1995, the Albanian population of Tetevo was demonstrating for a purely Albanian university. The Macedonian government was not willing to allow such an undertaking with the argument that this would further fragment

the country along ethnic lines. During the demonstration, police were attacked, and they shot one Albanian man. He was rumoured to have been from Kosovo. Some people argue that there is strong external provocation from Kosovo to divide the Republic of Macedonia with the aim to declare an Albanian state that would include Albania, Kosovo, and Western Macedonia.

3. For examples of regional studies, see Cole 1991; Hann 1993; Kideckel 1993; Verdery 1991.
4. See Kideckel 1995: 1; Hann 1993: 42; Berend 1986; Cole 1991: 14; Walzer 1992; and, for use of the term "ethnic cleansing," Denich 1997 and Todorova 1997.
5. See Pine 1996: 134.
6. Many girls had chosen to go to a language high school and to study engineering at university, whereas boys usually went to the technical high schools.
7. In these cases it is argued that there would not be enough money to pay even the bills—certainly true, in part. The suggestion that friends could move in together and use household articles that were being saved for their dowries to establish their household was deemed inappropriate. For young women to live by themselves was not acceptable if it could be avoided, which meant if they had family in Skopje. There were cases of young women living by themselves, but these were students from other towns who were studying in Skopje. This situation was accepted, though often commented on negatively by others.
8. Trouble would mean to get pregnant or lose one's reputation: "running around on the streets like a prostitute," said a friend's mother in 1994.
9. She paid the money back in the following months.
10. He is referring to buying and selling CDs legally and not dealing in illegally burnt CDs, which consumers would buy on the green market or in some new stores that import such CDs from Bulgaria.
11. My informant is saying this because Macedonian public opinion was that most money in the country was not taxed because it was earned through illegal activities, through smuggling with Serbia, through smuggling drugs, or through dirty deals in the privatisation process. Probably none of the private companies paid taxes, nor were "rich" people thought to do so since both had the connections to evade taxation. People said that only the honest and "small" people actually paid.
12. "Positive discrimination" is the term for a policy enacted to guarantee equal employment opportunity for the Albanian population, as many Albanians did not have the *vrski* (connections) needed to find a job.
13. While she was filling out job application forms, she was also filling out an application for immigration to Canada.
14. See Bringa 1995: 58ff for a closer discussion.

Chapter Five

Shopping for the "New" Person

Reflecting on the last fifteen years in Europe, people speak about how times have changed. Within this reflection there is more than meets the eye. Not only times have changed, but also time itself. Having waited for hours in line to buy small items in a "superstore" in Skopje in the eighties, I can comment on the way time has changed in the city. In the late nineties, I was actually able to go swiftly into a store and leave with my purchase in no time at all, at least in some stores, but I will get to that. There are different ways of thinking about and experiencing time: we assume linear time without acknowledging that we always live in the circular time of changing seasons, for example. And, we have to admit that, even though we aim towards some goal in the future, we also always expect that a good many things will continue in the future as they were last year. Obviously, Yugoslavia's not existing as it used to for my friends poses some dramatic challenges to their entire concept of time. No seaside visits in the summer on the Croatian coast, no winter skiing in Slovenia, and no autumn time spent in Belgrade visiting relatives or hearing a concert. As trivial as these things might seem, especially with the dramatic events starting in Bosnia and under threat of KLA infiltration into Macedonian territory, these changes effected my friends most dramatically.

Within this changed world, my friends experience new perceptions and interactions, and these come in a well-packaged form with marketed content. The whole process of transformation in Macedonia is, for the younger generation, largely defined by the introduction of consumer culture. This consumer culture influences their relationship to their bodies (which I will discuss in the next chapter) as much as it influences their relationship to objects. What this consumer culture then promises, in effect, is a future. But in a twist of time, my friends find it crucial to live out this future today. Changing times newly emphasise display and the external. Because the things once done in particular seasons may now not be possible, trust must be placed in things that are more predictable, and something very different has to be created, something that *does* promise to return next year. For my friends, these changes mean that, to a great extent, their understanding of themselves is enacted through their relationship to Western objects. It is a silent relationship in that it is defined as "normal," as normal as in any other Western country. Speaking of Estonia, Rausing notes the following:

> The dissolution of the Soviet Union means that the national discourse of future goals has shifted from a Utopian state, to a Western-identified "normality." The confines of being defined as Western within the Soviet context, however, means that the changes in material culture are greeted with less of the surprise, enthusiasm, or confusion, than might be expected: the "normal" or optimal reaction to the new things is a silent appropriation, redefining the objects as already taken for granted. (1998: 190)

My friends, in identifying Western goods as normal and not exotic, appropriate those objects in the same manner as the Estonians appropriate Western goods, as something that "naturally" belongs to them. Interestingly, this natural relationship for them derives, not from a historical connection to Western Europe, as, for example, Slovenia or Croatia could claim, but through claims derived from their Yugoslav past as a "Western" socialist country. In some ways this mixed past causes a perplexing contrast of thought in which Macedonia cannot be readily compared to other post-socialist countries, certainly not in the eyes of the people I came to know. For my friends, Yugoslavia would have lead "naturally" to the integration of Macedonia with the Western world. With the disintegration of Yugoslavia this "right" is seen as being questioned by the European community, which today demands visas for Macedonian citizens even though it once granted them free access. Nevertheless, consumerism gives the young people in Skopje access to "Western" goods and their appropriation as "normal" consequently gains a high political importance. It is not the access to "Western" consumer goods, but the recontextualisation of these objects[1] that allows my young friends to develop a very specific meaning, one that has evolved from their particular memories and aspirations. Just as a person's movement around the city determines that person's place in society,[2] a person is also determined by the consumer objects she or he purchases and uses.

What value do these objects have for my friends? Simmel insists that value is never an inherent property of objects, but is instead a judgement about objects by subjects, and that these objects are valuable because they resist our desire to possess them (1978: 67–73). Consequently, I would like to follow Appadurai's suggestion that looking at consumption should focus not only on sending social messages but also on receiving them and that demand thus conceals two different relationships between consumption and production. Demand is determined by social and economic forces just as it is manipulating, within limits, these social and economic forces (Appadurai 1986: 31). As my friends are receiving the social messages of the new consumer goods imported to Macedonia, the demand for these objects creates a new aspect of their identities for them, one that has social and economic implications.

Consumer objects bring with them images, stories, and ways of living from "outside." Sometimes these images become more real than "real life" in Macedonia; these images are creating a reality, creating a time in which to live. Often my friends referred to the world outside Macedonia as what "should be" and the world in Macedonia as a circus mirror that distorts. These consumer objects and their associated images become directly linked to people's lives. The criterion of what matters has become extremely volatile, and, as a result, people are left with an uncomfortable ambivalence in directing their own actions. Humphrey (1995), in discussing the consumer culture in Moscow, points out the uncertainties that may make it virtually impossible for her informants to know what they should or should not care about. In the same manner, Rausing's Estonians struggle to construct themselves as "naturally" Western in order to guard themselves from the ambivalence between their Soviet identity and their wish to be recognised as "Western." In Skopje, people's relationship to their surroundings and to the people around them changes slowly as the direction they take is towards the images of Western television. Friction arises if these images do not coincide with life as they experience it. The lives of my friends' parents do not correspond to these images, which causes friction between the real and an imagined world, which, in turn, causes turbulence in Macedonian society and the uncertainty that results in transformation.

However, to say that a society is undergoing a transformation only hints at the enormity of the changes this process brings to the individual lives of the people in that society. This transformation will alter personal relationships with one's family, employer, and even with strangers. The transformation began with the start of Macedonian independence and proceeded through Macedonia's break from socialist Yugoslavia and its embrace of what people understand as new, modern, and Western. Visibility causes these changes: the images on television, their enactment in the world of Skopje through the purchase of consumer goods, and the emphasis on a "body-beautiful" culture. The rearrangement of such visibility consequently causes a heightened sense of how appearance and self must be created and cultivated. To create a "self," my friends are forced to create social relationships that conform to their ideals and dreams, thereby creating "new time."

This "new time" gave a new meaning to consumer goods in Macedonia. There are three categories of goods. First are goods that can be received through kinship from the countryside; second are household goods that are received through kinship and friendship with people in Germany, Italy, or France; and third are goods that can be purchased through monetary means. I suggest that the last category creates something beyond *vrski* (connections); it creates an independent "Western" ideal that stands in direct opposition to *vrski*. This "Western" world is a world of achievement. However, this world is very difficult to imitate given the economic and political situation of Macedonia, which causes my young friends to depend more than ever on the family and *vrski*. An additional difficulty for my friends is the impossibility of finding independent housing. Getting mortgages from banks is impossible because they do not offer any of the usual services such as loans or even bank accounts. Bank accounts were closed in 1991; all savings were frozen and only exist as numbers on pieces of paper. Most people store their money, bundles of *Deutschmarks*, in their homes, hidden in an obscure cupboard. When the German bank changed its banknotes and I brought the new money back to Skopje, the mother of a friend fell crying on a chair, shattered because she believed that all the money she had saved, every *dinar* of which she had changed into *marks*, was now worthless. It took some time to reassure her that her money was not lost again.

In this situation, the modern way of delaying marriage until one is financially secure is obsolete and meaningless. The desire for independence and individual preferences cannot be considered. It is very difficult for young Macedonians, despite the intensity of their desire, to emulate the lifestyles they see in American television programs, which show young people leaving their parents to start independent lives. Only if their parents are wealthy, have some family property, or belong to the new elite of *biznis joveks* (businessmen) can a son or daughter move out. If they are most fortunate, they may move to one of the nice apartments in the newly built houses that are appearing everywhere on the Skopje skyline.

Most young people, however, must stay at home, submit to their roles as sons or daughters, and hope to find wives or husbands with their own flats. If not, a woman, for example, might find herself moving straight into her husband's parents' house as their daughter-in-law. All of my friends found this situation in direct opposition to the image they aspired to, the image they saw as modern and Western European.[3] Many times I was asked about my personal situation, and, when I described my independence from my parents, I was told that I could not possibly understand the situation they were in as my parents were not constantly telling me what to do and what not to do. This situation, close family links and obedience to one's parents, is common in many societies, but my friends rejected this way of living. They had thought that this way of life, which they associated with their grandparents, was passé, at least in the urban environment of Skopje unless you were Albanian. My friends believed that this way of life was not meant for them, and their desire to recreate themselves as modern citizens of an independent Macedonia created pressure at home, which they sought to escape and possibly negotiate through their love of foreign consumer goods.

There is a distinction between "liking nice things" and the meaning that is applied to those "things." To like and want to have a sweater is one thing; to want it so much that you are prepared to spend a month's salary on it, is another. Not all my friends spent money in this way all the time, but they certainly all felt that this spending was acceptable. When I asked about specific purchases, I would often be told that, unlike them, I could easily buy the item in Germany. The explanation given for this inequity would be that there is greater freedom in Germany; it would not be said that Germany is simply a far richer country. Their idea of "freedom" was contrasted to the lack of freedom they experienced at home where they were not able to do what they wanted to do. What they wanted was to be like "everybody else," that is, like people their own age in England, Germany or America. As I was told many times by numerous people, "All I want is a decent life." They feel their difference from the young people they see on Western European television. On television, young people of their age live by themselves or with friends, manage their own households, have jobs they chose and applied for, and have hobbies and a "lifestyle." They have their own money with which they can buy whatever they want. They travel and know many interesting people and places. My friends, however, are constantly scared to lose what they have: peace, money through the devaluation of currency, ideals, and dreams. Buying expensive consumer goods today gives them a feeling of security about tomorrow. After all, by tomorrow those goods might be out of reach forever. So their shopping gains a certain sense of urgency. This urgency affects not only my young friends, but also their parents. Like their children, their parents want to see their children have "a decent life." Their parents see the good life as being achieved through consumption. The parents themselves, however, rarely buy consumer goods for their children, nor are they necessarily willing or able to deplete household resources in order to buy anything that is beyond necessity. They would, for example, buy a winter coat for their daughter, but would not be willing to buy a "Western" designer coat. The daughter, however, might add her salary to the money her parents put aside in order to buy the coat she wants.

The television images from Western Europe and North America support my friends in their quest to control their lives; they want to become those images in order to escape to a different world. This world should not be understood as simply an illusion because it indeed exists. It creates a particular reality, a reality that presents my friends with a mixture of old Yugoslavian values given new meaning through Western consumer goods and images from which my friends create their personhood.

In the years following the fracturing of Yugoslavia, the appearance and practices of people in the street and, accordingly, the appearance and practices of my friends have changed, while social categories have not. My friends describe this situation:

> … it is hard to express these thoughts; it seems things changed, but words have not. I do not know what is right or what seems wrong. My parents know things, I know them too, but they do not fit with what is today, but then there is nothing else either, it is not like we have different ideas on things. My words do not express what I think anymore.

In this time of uncertainty, where words are not adequate to convey thoughts, objects and their consumption provide the sought-after guidance. In this world, neither ideology nor ideas, but material objects represent my friends' dreams and fears. The valued perfume, sweater, or brassiere with an English, French, or American label provides guidance in the search for defining oneself beyond the limited definition of being a daughter, beyond village, beyond being an engineer, and beyond being Macedonian. The consumerism of my friends functions in a symbolic replication of European consumption thereby promising the arrival of "Europeanness." The goods that symbolise this "Europeanness" can be seen as symbols of specific aspects of European life: prosperity, power, and "happiness." Consumption in Skopje, then, creates consumer objects that are not defined through their utility within a global exchange but through the role they play in the symbolic system of identity formation.

Consideration of the issue of food, its consumption, and the ambivalence of the younger generation's social practices provide insights. This generation is altering the social order within the city. On a visit to Berevo, where I stayed with grandmother Baba Mare, I began to understand the meaning of the food, which I was repeatedly offered on my visits to Macedonian homes. At breakfast, when I ate a combination of eggs, eggplants, garlic, peppers, and tomatoes, Baba Mare pinched my cheeks and said, not so much to me but to the other members of the family, "She is Macedonian. She eats peppers and tomatoes with us, she is one of us."

"One of us" was a much used and highly significant expression for Macedonians in relation to "outsiders." One of the first questions asked of the person introducing me was always "Is she one of us?" My host would reply, "No, she is not one of us." On my first visits to people's homes I would, with formality, be offered *slatko*, a sweet usually made of young figs, apricots, strawberries, or other fruit cooked many times in sugar. Only morsels of *slatko* were eaten at any one time and were followed by a big gulp of water because *slatko* is extremely sweet. Offering *slatko* is a traditional Macedonian custom and a sign of great hospitality.[4] It is also a source of immense pride for each *domacinka*, female head of the household, who prepares her own *slatko*. Nevertheless, I was told that offering *slatko* is a Turkish custom, Turkey being the source of many "Macedonian" customs, and as such, offering *slatko* is a constant reminder of 500 years of "Ottoman oppression."[5] These old traditions, then, have two meanings: they define "who you are," but they also identify past oppression. My young friends, although still very proud of their mothers' *slatko*, feel that an independent Macedonia needs different traditions, so now a formal visitor often receives sweets bought in the duty-free shop.

When I began to eat regularly in a household, and thus, to regularly eat the staple dish of peppers and tomatoes, I became more like a "real" Macedonian. The classic pepper and tomato dish is *aivar*, which is made at harvest time. *Aivar* has come to represent the extended family to the people of urban Skopje who gather once a year to cook it. In jokes referring to the possibility of a Serbian invasion, people loved the idea of every single Macedonian standing up for his or her country and, because lacking any weapons, throwing tomatoes at the Serbian army. Tomatoes and peppers are the staple ingredients in Macedonia today and are, in fact, the country's lifeline. Besides watermelons, grapes, poppies, tobacco, and some fruit, agricultural production in Macedonia is based on peppers and tomatoes. The dismantling of Yugoslavia left Macedonia with a very one-sided economy, a situation that, if it is to be changed, will require revolutionary measures.

Rakia (schnapps) is another strongly codified medium of Macedonian social exchange, the consumption of which turns out to be a statement about gender. Just as the *domacinka* is proud of her homemade *slatko* and the whole family takes pride in its *aivar*, the *domacin*, the male host, takes pride in his homemade *rakia*, made from grapes, plums, or potatoes. *Rakia*, "cleanser of the soul," is a very strong alcohol and is seen as a man's drink—it is never offered to visiting women. Wine can be consumed by both sexes, the sweeter ones suitable for women. Today however, the idea of sweetness is associated with excess, indulgence, and Ottoman rule: very sweet treats such as Turkish delight or *slatko* are associated with the Turkish occupation. Instead of consuming sugar, fashionable young women today follow Western patterns by neglecting "opulence" and using "sweetener" in coffee. Women also rarely drink beer. These gender divides are more rigid in the countryside, where it sometimes is deemed unacceptable for a woman to drink any alcohol except wine. Traditional food and drink create and recreate traditional social relationships in a way that is largely implicit for the participants. The introduction of imported consumer goods, because they arrive unencumbered by tradition, therefore represent a contrast to the social relationships produced by the consumption of traditional food.

My friends in Skopje, who seek a "more European style of living," consciously alter social norms for the consumption of food. Even though they understand tomatoes and peppers, *slatko*, *rakia*, and coffee as being important, my friends insist on the past tense when referring to them, and they consider these traditional foods and drinks as representative of their childhood, their parents' and their grandparents' generations, and of Yugoslavia. Gell stresses the same intense relationship between items consumed and the social identity of the consumer that I wish to stress for my informants: "What distinguishes consumption from exchange is not that consumption has a physiological dimension that exchange lacks, but that consumption involves the incorporation of the consumed item into the personal and social identity of the consumer" (Gell 1986: 112). *Slatko*, *rakia*, and other traditional foods and drinks are still consumed in Macedonia, but "Western'" consumer goods, food, material objects, and images transcend the merely utilitarian aspect of consumption goods, so that they become something more like works of art, charged with personal expression (Gell 1986: 114).

In 1988, I spent many long evenings with a group of female friends in Skopje's cafés, drinking Jupi or Cokta instead of Sprite or Coca-Cola which, imported from Greece, was only available in a few places. We drank Turkish coffee or Nescafe. Today, we order cappuccino, espresso, and banana milkshakes. In 1988, we spent many hours at each other's homes eating *kashkaval* (a hard cheese) and chickpeas and having the coffee grains in our cups "read" by male friends who would look deeply into our eyes when they told us our futures. Or we would simply indulge in straightforward *muhabeti*, the Turkish word for socialising and talking sociably.[6] In 1996, we would stand outside Van Gogh listening to loud music, drinking cocktails and "meeting" people. Such are the changed times.

In Macedonia, food, what type you eat and how and when you eat it, defines your gender, the level of formality in your relationship, and your age: it communicates origin and identity. In this context, the change of diet undertaken by many of my friends becomes highly significant because they are consciously attempting to change their image of themselves through changes in their eating habits. Macedonia has established an interesting compromise, in terms of its identity, between its past as a part of Yugoslavia and its present as an independent country. In the midst of massive political upheavals, my friends are trying to make sense of the constantly changing and conflicting images of what has been, what has come, and what is coming.

EUROPE AND AMERICA AS IDEALS

Western television and its associated images have become dominant in Macedonia since the fall of Yugoslavia and are now central to family life. The increasing openness in Macedonia since the fall of communism has lead to a huge increase in the availability of pornography, which, in turn, has clearly lead to changes in people's perception of themselves.[7] Pornography is easily available to anyone, including children who can watch pornographic films at three o'clock in the afternoon. Although pornography is considered Western, images of naked women were already found on crossword magazines in Yugoslavia despite party ideology. Today, however, as pornographic films pirated from American satellite programs can be seen at local cinemas and on television, I suggest that the meaning of these images is deeper than the one immediately assumed. Catherine Portuges refers to pornography in Hungary: "In these and other works in which youthful bodies are exhibitionistically fetishised, the ardently sought free-market economy is both symptom and cause" (Portuges 1992: 287). For many of my male and female informants, however, images of naked bodies do not so much present women as objects, a point not easily understood by either male or female informants, but represent a meaning that stands in connection with the act of obtaining those films: pirating.

The act of pirating films from a satellite dish means gaining access to an objectified area of the world that is "out there" and unreachable other than through "pirating": the "have-nots" stealing from the "haves." The women in the films are stereotypical American women by Macedonian standards; these women represent America. To connect to this world is to consume its images in a manner assumed to be "Western": detached consumption that objectifies everything, only taking. Watching pornography is more than a symptom and aspect of a market economy, more than entertainment; watching pornographic movies is an act of gaining access not to the object of the film itself but to the act of watching it. While I lived in Skopje, male friends came to visit me, not for a coffee and chat as I thought, but in order to tune in to the different pornographic movies that were playing. My loud protestations were heeded but not understood. They did not understand why I was upset, as their understanding was that I was from Germany and porn was a normal thing for me to watch. Also, I was their friend, so why was I not ready to share the freedom of living on my own and my television, when they surely could not watch such films with their parents and little sister at home. I had two things they saw as "Western": choice and freedom.

By recreating themselves as modern consumers, my friends attempt to eradicate the differences between themselves and the West, differences that have become increasingly skewed. Having been raised to be proud of their Yugoslavian identity, Macedonians, citizens of a non-aligned nation country that is no longer a "model" or the Switzerland of Eastern Europe, find themselves disenfranchised.

REACTION TO MACEDONIAN INDEPENDENCE AND THE FALL OF YUGOSLAVIA

Socialist ideology has lost its importance since the fall of Yugoslavia and, consequently, Macedonia has also lost its old fixation with programmed thought and social activity. In its place, a diffuse and diverse consumerism provides a political perspective. In the four years between 1992 and 1996, consumption became the key feature of the "new society" of Macedonia.

This culture of consumption is centred on the capital city of Skopje and developed at a time when many people in Macedonia had to struggle to survive. Through the process of privatisation, many people lost their jobs as companies closed down; they lost their small amount of foreign currency because bank accounts were frozen, and many people were barely surviving. Most people simply could not afford to buy foreign consumer products.

The development of a consumer culture results from a particular set of historical circumstances. How is this transformation seen from within? Consider the effects of this transformation as shown in an excerpt from my notebook in July 1996:

> The last old shopping centre is empty. A new mall, "Beverly Hills," has been built. In the old bookstore they opened a Doc Martens store. Things change—even if only slightly. Many foreigners are here, and Stefan, an engineer friend, was in Ohrid to install equipment for an action movie by Steven Spielberg Productions starring Nicole Kidman and George Clooney.
>
> They took down the old bridge over the Vardar and found that the base can be dated back to Roman times, 300 BC. This means that the Bridge is not a Turkish construction, a very convenient situation.
>
> Sitting here and writing this down on a sunny afternoon at café Ciao, it is absolutely obvious that Skopje today is a young people's city. The only people you see in the streets are between twenty-five and thirty. These are the people without their own families, maybe without jobs. One hardly sees any older people. If you were to look for them you would find them at the green market or in the park going for a stroll or playing Backgammon.
>
> Yesterday we all went to café Van Gogh, which is now closing at 12 p.m. Everyone was there. Suse still has five exams to go; Elena, who works for a Swiss company, is very well paid but has to work from 8 a.m. to 9 p.m. She is designing software programs for the post office and banks. Zoran came too; he looked so happy. He is getting married. I was talking to Igor; he still has to finish one exam. He is working as a programmer too. He was telling me that everybody wants to leave Macedonia, that nobody wants to stay here. He was saying that his mother really surprised him when she told him to leave. Zoran still cannot find work. I asked him if there was any business going on, but he said that the times are bad for businesses, it is not the right time.
>
> In the morning we went to have me registered at the police station. Afterwards we went to Eva's place; she works for a private import/export company. She was complaining that there was a lot of work to do. I talked to her colleague Daniela. She told me what a beautiful country Germany is, and we ended up talking about the different perceptions of work in the East and West, and she told me that Macedonians are not used to work. She said everybody wants to go back to the days when one finished school and automatically got a job, which earned you enough money to live an okay life and to travel. She said that everybody misses Yugoslavia, that it was a good life.

If my friends are working, they are usually employed by the state or by a private company. When they are employed in a private company, their work is always related to or directed towards Western Europe, the UN, or an American aid organisation. If they work for the state, they complain about not having enough work, not having enough pay, and about widespread inefficiency. All of them, however, embrace the "New Macedonia" and its revised interpretations of past times and capitalism.

The lives of my friends in 1996 are bound up with the discussion of the "new people," those who are envied because they can afford the consumer culture but whom my friends judge as stealing from the people. None of my informants would count themselves in this category although they certainly aspired to the same consumption patterns. The idea that making money is a form of theft is, as Liseta pointed out to me, an old concept from communist times, when making money for oneself by securing a deal for one's company was seen as "stealing from the people." Today, she commented, it is called good business. In fact, the old ideas live alongside the new. The result of these conflicting ideologies, the Yugoslavian ideology versus capitalism, which is the true essence of independent Macedonia, is evident in the constant talk of corruption. To steal is "to be corrupt," but in today's Macedonia, stealing could just mean good business. Rausing reports a similar concept in Estonia, where, despite the government's commitment to the free market, the people who have been successful in it are often seen as less "Estonian" in people's imagination—as very smart and with dangerous connections. As she says, "They seem already to belong to another imaginary entity that is only partially contained by the entity of Estonia" (Rausing 1998: 195). This "imaginary entity" symbolises for Rausing's Estonians what the "Western European" signifies for Macedonians: a specific type of person who has success through achievement and not through intelligent manipulation. The latter characteristic is seen as an original Balkan trait. The nouveau riche represents the typical Balkan personality, and, in addition, they do not conform to the existing socialist ideology because this class is seen as stealing from the common people. This explains the lack of a drastic emergence of the nouveau riche as a class in itself as Humphrey (1995) reports for Moscow.

For many of my friends, their parents, and relatives, the knowledge and experience they have acquired has become meaningless in the world of "business." In this world, a polarisation takes place that stands against socialist and "European" valued ideology, as both are seen as valuing equality in some way. The middle class as "Western" ideal disappeared in Skopje before it was formed. Possibilities increased and the standard of living rose for a few people, but for most, it declined. For my friends, the struggle of who they want to be and who they can be takes place in this world.

Shopping is the locus in which this particular struggle becomes visible. Western television conveys an image of the world where supermarkets predominate, where one can get the things one needs within minutes.

Life in Skopje looks very different however; my friends disregard this disjunction and interact with the outside world as if it were actually available to them. It follows that economic activity has intrinsic political and ideological value. If my friends browse through shops in which they will never be able to afford to buy something, they are, nevertheless, making a political statement. This intrinsic political statement is determined by history, in this case, the history of socialist Yugoslavia, as well as by Western television and global consumerism. In socialist Yugoslavia, production was glorified, whereas on Western television, wealth produced through speculation is glamorised. Each value condemns the other, but promises the same thing. In Skopje, the world of consumerism arrived because of risky speculation, and the contact this produced with the outside world embodies a particular political/moral attitude. Shopping in Skopje illustrates these points well.

SHOPPING IN SKOPJE

Macedonia aligned itself with the material and ideological culture of the West, a process that began in the mid-1980s when Macedonia was part of Yugoslavia. "The West" was understood by everyone as West Germany and America followed by the more intellectual choice of England and France, which were only important to people who had travelled there. Because of migrant work in Germany by at least one family member and the legal or illegal supply of German goods, Germany was the main supplier of Western consumerism, both ideologically and practically. Ideologically, Germany was closely followed by American television. Interestingly, the green markets were the centre of German consumer goods distribution and have given way to high fashion shops like Orka Sport, Yukan, Stefanel, and Benetton, a situation resulting from an ethnic shift as well. The green market used to be run by ethnic Albanian Macedonians who had been guest workers in Germany. They brought German consumer goods back to Macedonia and sold them, for the most part, legally in the green markets. Today, the new shops are run by the new Macedonian business elite that travels to Germany, acquires similar goods, usually legally, and sells them at three times the price the Albanians charged. Many import-export agencies that play on the idea that the goods they offer are as desirable to German customers as they are to Macedonians have sprung up around Skopje. In contrast, the green markets are thought to offer goods considered cheap in Germany. In the socialist Republic of Macedonia, basic goods were found in state-owned shops or in semi-legal markets. The markets supplied the population with farm products from outside of Skopje as well as with goods illegally smuggled from Turkey, Greece, and Germany.

Goods that were scarce were obtained through friends or connections at the workplace. Supply was never assured, even of everyday products, and items could disappear from the shops overnight. Then, one's network of friends would provide advice as to which shop might have toothpaste, sanitary pads, toilet paper, noodles, mushrooms, or other goods. In 1992, small private market shops began to appear slowly, from which people sought to make their fortunes. Many failed, but some people managed to build empires such as Orka Sport with the help of "old" illegal money.

Stores having contacts with Turkish, German, or Greek companies despite the Greek embargo, as was the case with Orka Sport, could obtain the Western goods to attract customers. Duty-free shops were established and could be found anywhere in the city as well as on the borders with Serbia, Albania, and Bulgaria, although these shops have since been outlawed. These shops sold traditional duty-free goods such as Marlboro cigarettes, Metaxa, and Braun watches. As time went on, however, they sold refrigerators, microwaves, Gillette shavers and blades, vacuum cleaners, batteries, and Ariel washing powder, as well as Chanel perfume and other luxury items. These duty-free shops turned into shops selling common consumer goods. New shopping malls were built, and many tiny shops of around five square metres sold lingerie, sweaters, skirts, watches and other goods that could readily be identified as "Western." Some of the shops expanded and became Levi's stores, Stefanel stores, or, in one case, the Yukan Department Store, in which the customer was offered cake and coffee with Toys 'R' Us toys and Stefanel and Hugo Boss dresses and suits for him and her.

Shops that I would call small department stores are often bulging with Western goods and employ a staff that is well dressed and well mannered. These shops are run by the typically nouveau riche, socialist managers who bought up their companies through the illegal devaluation of those companies or people who made a fortune through illegal import and export operations with Serbia, which was under an embargo by the European community. Many of my friends automatically cross to the other side of the street when they pass such shops, feeling it inappropriate to walk next to the "new" shops, as if the shops suggest a world to which they do not belong. Shopping, then, is still a long way from being a leisure activity in Skopje with its burning pavement and dust hanging over the city; this world is almost surreal when seen through its haze. The shopping depicted on television is still a long way from the reality of Skopje. Miller (1997: 39) describes shopping as a social occasion to look at other shoppers. In Skopje, however, the shopping malls seem relatively deserted while the green markets are buzzing with life. Shopping in Skopje is not a social occasion, but a direct encounter between the objects to be bought and the shopper who would like to identify herself with these objects, not so much to present herself to others through the object, but to stress a specific aspect of herself.

Shopping in Skopje is not so much about personal desire; it is a political enactment of choice. If the shopper can access the specific choice of buying "Western" consumer goods, such as Ray Ban sunglasses, she is partaking in a life like that of her fellow shoppers in Germany or England. The fact that her choice does not fit into the economic environment of Macedonia makes her choice a political act. In Skopje, friends rarely shop together; only the "expertise" of the "Western" and shopping-committed anthropologist was sought. Generally, shopping for "Western" consumer goods is a solitary enterprise. In contrast, shopping for food and necessities is often done with a parent or siblings. Friends only accompany the shopper if a present has to be selected for another friend, partially because presents are commonly given from several people. Spending money, then, cannot be seen as a demonstration of financial means, sociability, or sensual satisfaction as Miller (1997) suggests for shoppers in Trinidad, but has to be understood as a real and desired barrier between the shopper and "normal" life. However, Miller's comments sometimes coincide with mine for Macedonia:

> It is through consumption that political values may be formulated in such a manner that populations are at least able to appropriate the images of the possible worlds they seek to create. Through this objectification they come to consider who they might be in relation to the gamut of political identities (Miller 1997: 45).

The reasons that Western consumer goods are imported and exported are secondary to the primary issue of the specific symbolic value of Western objects in Skopje. This value is not created either by the "European" media or by the Macedonian government fulfilling its political agenda, but by the shopper in Skopje who actively chooses the symbolic value of the objects. This act of shopping is a solitary one, but the value ascribed to the objects that are shopped for by all my friends in today's Skopje lies in the idea of living a "normal," and therefore European, life.

In socialist Yugoslavia goods were not bought in an exercise of free will, but were allocated by the state or the companies. Thus, though people bought things, they felt that they were receiving something from the state. Shops were not called Orka Sport or Yukan, but by the more general nomenclature of meat, shoes, bread, and books. Shops did not always have window displays, and even if goods were displayed, what was displayed was not necessarily for sale. Even today, many of the small shops do little to present goods in an attractive manner, unless they wish to make them more expensive. The window displays are a mix of many products ranging from Italian noodles and German black-bread to shaving cream and diapers and an expensive suit. Today many shops carry only one of each item they display. Others stock just a few of each line, so it is important to buy an item when you see it. The only exceptions to this practice are the internationally owned stores such as Benetton, Benetton for Children, and Stefanel, which stock quantities of the same product. Many goods are priced in German *marks* and bought using German currency even though the *dinar* has been stable for a number of years.

The green market still flourishes because it is the only place where fresh fruits and vegetables, cheese, fish, and cheap Turkish household goods can be obtained. Imitation Levi's jeans are still found there, but now the UN peacekeeping troops are the only regular customers.

The state store offers one plate on ten otherwise empty shelves while the shop assistants sit in groups smoking cigarettes and drinking coffee, taking no interest in customers, just as they did in the old days. Bored and uninterested as the shop assistants are, one of my friends admitted that she feels more comfortable in the state shops than she does in the new shops where the assistants try to persuade you to buy. To say to a shop assistant that she only wants to browse feels like an admission of inferiority. My friends say that they feel they are not actually meant to buy these items, which are meant to be for non-existent glamour girls.

Not all of my friends were attracted to such shops; some dreamt of actually going to Germany or London to buy their favourite items. A friend of mine managed to travel to London and insisted that she would not spend her money senselessly while she was there. Nonetheless, on her return, she was proud of all the nice things she had found in London, and, judging from what she bought, goods were much cheaper there than in Skopje.

What is important to understand is that shopping in Skopje, even in 1996, had nothing in common with the experience, which they saw on television, of "going shopping" in London, Munich, or Paris. There was no relaxed strolling through shops; it was a hunt for something special, something Western that was quite inexpensive without being shoddy. People constantly feared that products would disappear without warning from the shops. When I described to a colleague in London the half-empty, half-finished shopping malls that rise high above the sky-line of Skopje and the shopping habits of my friends, he commented that these shopping malls made him think of cargo cults in the South Pacific. The malls indeed looked like they were planted there by some giant: bare concrete, stairways that are too long and broad, winding their way upwards toward nothing, broken glass windows, and hallways that are too long for the tiny shops they access.

If there was a desire to imitate Western consumerism, the old ideals prevailed over it. The way my friends "go shopping" appears to be a microcosm of political life in Skopje, in the same way that the bodies of my friends are one expression of the country's search for a new way to mark itself as Western. The process of buying cheese exemplifies this: a friend of mine caused a minor riot in one of the new grocery stores by refusing to take a portion of cheese because it had been cut too big and was, therefore, too expensive. She said that, a few years ago, she would have accepted whatever the saleswoman had given her; she would even have brought her own paper with which to wrap the cheese. No more, she said, explaining her current refusal in this way: "I do not see anybody in Germany or America doing this. Here in Macedonia, we only learn to accept, never to question, never to insist on something for us, for ourselves. I did not want to pay more for the cheese. I knew how much cheese I wanted and how much I wanted to pay." When she refused to accept the cheese, another customer in the shop supported her, insisting that it was my friend's right to choose what she wanted and that the other women in the shop should follow her example because she was a "new" woman. Here the concept of "newness" becomes political: newness means the right to demand what you want. The political discourse of the "new" in "new" Macedonia refers to a right not articulated in mind or speech when Macedonia was still a part of Yugoslavia, according to my informants: the right to choose. The concept of choice, of choosing, even demanding what rightfully belongs to any human being, is at the very heart of the image of Macedonia that Macedonians wish to convey to the rest of Europe. This "right to choose" is what Western television promotes all the time: consumer goods offer choice, and choice is a right that has to be granted in any modern state. "No choice, no freedom, you have Eastern Europe all over again," a friend said, trying to help me understand the Macedonian craving for German and Italian advertising programs on satellite television.

Many of my friends do not see themselves as having the right to demand or to choose and are actually taken aback by this attitude. Such an attitude sets you apart from your fellow humans. Even if they would like to participate in the "new" Macedonia, they grew up in Yugoslavia and are nostalgic for those times, for a "better and peaceful life, when people cared for each other." In this sense, my friends follow conflicting notions in their task of negotiating themselves as citizens of independent, "new" Macedonia. They feel attached to their inheritance from Yugoslavia, but they also want to be part of Western Europe. In the midst of this conflict, my friends try to make sense out of their lives.

Be it shopping or the type of body a woman is supposed to have, what is going on in Macedonia today is a fierce fight between a cherished legacy from the Yugoslavian era and what people believe is necessary to differentiate themselves from the Balkans and Eastern Europe. It is a conflict my friends have to put to rest for themselves. It forces them to think about who they are and who they want to be, and, in the process, it clarifies the social position of others and also societal norms. Each of my friends engages in a project of their own, be it dieting, bodybuilding, or the purchase of expensive clothes, appliances, or furniture. Each project helps them to redefine themselves. Nevertheless, in the current situation, my friends are trapped, some of them tragically, in a world of illusions.

POLITICS AND CONSUMERISM

Despite the seeming contradiction, many people had greater access to foreign consumer goods when Macedonia formed a part of communist Yugoslavia than they do today. People had relatives or friends in Germany or Italy; people travelled to Turkey, Italy, Germany, Spain, or England and obtained consumer goods abroad. From 1991 to 1995, access to Greece was denied because of a Greek embargo and, since the fall of Yugoslavia, hardly anyone is able to travel to Western countries because obtaining a visa is so difficult. One must work in an embassy or have excellent foreign connections to leave the country.

What drives the desire to be a new person, to be a new country? Certainly there is the missionary zeal of the Western industries that have started to invest in Macedonia in order to bring a post-communist country towards a free-market economy. Today you find Western ice cream everywhere. The old Italian style ice-cream shops have lost many customers, even though one scoop of their ice cream costs about five cents and a Magnum or Snickers ice cream costs two dollars. Worlds drift apart here. Parents struggle to make ends meet; some going for months without pay because workplaces are basically shut down.[8] Adult children spend the money they earn on consumption goods instead of contributing to their parent's household, which is how the world should be according to both parents and children. "The world closes down on the children, but what is offered now, as long as the West is still noticing Macedonia, the children should enjoy. Who knows what will happen tomorrow?" A mother made this comment while discussing her children and their desire for expensive perfumes, clothes, and cars.

In the nineties, although essentially still a socialist country, Macedonia embraces the concept of a "new" Macedonia, but does not copy Western capitalism as a political system; thereby, Macedonia essentially creates a new model for its people. A Macedonian judge explained it to me:

> Macedonia wants to be European; it wants to be a modern democracy, but we do not want to turn our back on our past as part of Yugoslavia. Yugoslavia has given us more than an identity, as Greece always likes to point out. It has given us something to believe in, and we do still believe in it. Macedonia has historically always been stepped on; only within Yugoslavia were we given respect as people. We believe that everybody should have equal rights, and we do not like the American idea of "the best only should win." We, as a whole people, want to win in this enterprise we call the Republic of Macedonia.

The young people in Skopje reject the timeless, unchanging aspect that goods possessed in socialist Yugoslavia. They long to be young, ambitious Europeans. The limitations of choice are felt more strongly today than before because this limitation goes hand in hand with restrictions in movement. In the past, I did not hear complaints that one flat looked much the same as another because the Yugoslav state supplied similar furniture to everyone. Now, when my friends think of establishing a home with their husbands, they dream of new Western furniture. However, obtaining such furniture in Macedonia is impossible, and, even if it were available, it would not be affordable. Today, many of my friends are happy if they can afford new furniture at all, even Yugoslavian-style.

While Macedonia was part of Yugoslavia, my informants considered access to the necessary private "consumer" goods a right, whereas today consumer goods have become a component of one's personality. They define whether one belongs to the "hard working, but going nowhere" or the "fashionable, should be part of Europe" group. In newly independent Macedonia, consumption patterns determine whether one is defined as "new," and the "new" in the country's political-social discourse is embraced. Macedonia itself, its neighbours, and Western aid agencies expect that Macedonia will reject its socialist past and choose instead to be a capitalist democracy. To be recognised by the European Community as an independent country and not as simply one of the leftovers of the dissolution of Yugoslavia, Macedonia has to convince the EU that its only goal is democracy.

My friends were devastated when they were denied access to Europe upon applying for visitors' visas or for acceptance to international conferences; they had felt that their desire was so strong that it must be fulfilled. This denial of access led my friends to get to the West by other means: the consumption of Western goods. "If I cannot go to Italy, I will at least buy a Benetton sweater here," a friend said, explaining that she wants to be worth that much. However, this consumption is not understood as a rejection of socialism; rather, Western consumer goods have become a way of constituting one's self and rejecting the barriers that have been built between Macedonia and Western Europe. If consumerism in Macedonia is understood in this way, then the situation is markedly different from that described by Verdery and Humphrey.[9] Verdery (1995: 7) states that, in Romania, "the present confrontation between capitalist and non-capitalist systems is being played out as a cosmic struggle between Good and Evil." I argue, however, that, in Macedonia, the meaning of consumer goods has changed. The meaning is not determined by an assignment of "good" and "evil" or by the meanings of the socialist past and post-socialist present. In the mix of old and new, consumer goods are used by my friends to help them create and define themselves as European, thereby creating and defining a new timeline. Through the acquisition of Western consumer goods, my friends construct a sense of being "deserving" of a "Western lifestyle." They do not try to divest themselves of their Yugoslavian past. However, the contrast between their Yugoslavian past and their present position[10] creates a desire to incorporate Western consumer goods, images, and ideas into their Yugoslavian identity. In their consumption, my friends struggle to retain a sense of themselves as modern urbanites, a sense that was created in socialist Macedonia, while incorporating something of the freedom and access that is seen as "European." Thus my friends' consumption is a key element in creating their idea of an independent Macedonia, a place of urbanisation and modernity (Miller 1987) combined with freedom and access. My friends' consumption, though not an affirmation of socialism, does not oppose the old political system either. What my friends wish is to be "Europeans" without stripping socialist Macedonia of its meaning. It is this image unanticipated by my friends' parents, the politicians, and the aid organisations that steers Macedonia on a strict course towards Western capitalist democracy.

THE OLD AND THE NEW VALUES

The concept of "price" within Macedonia is notable. During the socialist period the cost of most necessities, like bread, were subsidised while other prices, such as the price of flats, were fixed by the government because in some sense housing or appliances or furniture were owned by the government and thus the people.

With independence came privatisation, and flats were transferred from the government to private owners, allowing many people to profit greatly by moving to the countryside or in with relatives and renting their flats to UN personnel and the peacekeeping forces. Because of privatisation and increasing demand, the cost of renting a flat has risen sharply: one can spend up to $500 a month. Only foreigners can really afford to pay such rents. Nevertheless, since the 1990s, the construction industry has been booming in Skopje, and high-priced consumer goods are found in several "modern" and "new" shops.

Apparently, the price of consumer goods is not perceived as related to the cost of production. My friends and their families are uncertain as to the value of consumer goods; they do not know an appropriate price, and they cannot compare specific items to similar things in the West, except by decoding Western advertising programs. However, value derived from Western advertisements carries its own ideology, and my friends understand value as a price tag on "identity." Within this logic, a "new identity" necessarily implies that objects bought to acquire it were not cheap, but rather had a "Western price," the price of Western-ness. Many people in Macedonia asked me what I had bought each day, as they assumed that everything in Macedonia, including the "new" luxury items, must have seemed cheap to me. When I told them that I had not bought anything, they were puzzled, wrongly assuming that my lack of consumption was because I was not interested in anything available in Macedonia, rather than the real reason: I simply could not afford to buy goods there. This explains my friend's amazement upon returning from London, where she realized that many things were cheaper there than in Skopje. Another myth, that in Turkey consumer goods are incredibly cheap, creates its own truth. People travelling to Turkey will often pay whatever is asked because they do not know what a fair price is, and are sure that they must be getting a bargain because they are in Turkey.

The concept that price should be related to purpose and the benefit brought by a product is understood in relation to products that were available in socialist Macedonia, but not in relation to goods carrying a specifically "Western" value. How is this value created? The value of "Western" goods is determined by demand (consumption) and desire. As Appadurai (1986: 29) discusses, demand emerges as a function of a variety of social practices and classifications, rather than as a mysterious emanation of human needs, a mechanical response to social manipulation, or the narrowing of a universal and voracious desire for objects to what objects are available. Demand and, therefore, consumption are aspects of the overall political economy of societies. Hence, consumption is eminently social, relational, and active. A new value, for example, was created by assigning value to living expenses, whereas previously, in Yugoslavian times, living expenses had had no monetary value, and accommodations were instead seen as a right to which everyone was entitled. When it costs $500 per month to rent a flat, a price that few in Macedonia are able to afford, the high cost of living creates an idea of the "new" lifestyle of independent Macedonia. Some Macedonians live in these expensive flats, and these people are defined as "new," as the only ones capable of living this new life. Consequently, most people in Skopje are strangers to this lifestyle.

If we consider consumption and the value of goods and services in Skopje, we have to focus on consumption not only as sending social messages but as receiving them as well. Demand then, as Appadurai (1986: 31) details, is determined by social and economic forces, but can also manipulate those very forces. I suggest that Western consumer goods in Macedonia are regarded not as luxury goods that simply fulfil a desire, but as an index of social value that is ultimately political. Consequently, the value of "Western" goods in Macedonia is not determined through demand for these objects, as in the West: many "Western" goods are in very low demand in Macedonia, and, often, only one item of a particular kind could be found in Skopje at a time. The value of goods was determined by the political positioning of Macedonia as a "lower kind" of Europe. The prevalent thinking was that goods must be cheap for me in Macedonia because I am truly "European," and my choice not to buy anything in Macedonia was made because I assign the right political value to Macedonia. My friends interpreted my lack of consumption as my seeing Macedonia at "a lower level than Europe." It is inconceivable that "Western" goods in Macedonia are actually more expensive than in most European countries where demand, for the most part, determines the value of goods.[11] In Macedonia, the value given to "Western" goods is that which is seen as necessary to lift the nation to "Western" standards. This assignation of value is not done by imitating "Western" prices, but by aligning the value of goods to a scale of "uplifting." The value of these goods is determined by the believed difference of value between Macedonia and Western Europe.

THE NOUVEAU RICHE

What is meant by the term "the nouveau riche?" A negative connotation originating in socialist ideology suggests that the people of this new social class are stealing from the country because they make profit without investing their own labour in the process. They are like traders, condemned by socialism, who buy and sell goods without adding any value through their own labour but still gaining profit from their transactions. Characteristically, in 1992, illegal trading was booming, especially because of the embargoes on Serbia by the EU and the UN and the Greek-imposed embargo on trade with Macedonia. Macedonians sold petrol to Albanians in Kosovo and Serbia; Macedonian and Greek authorities smuggled whole trains filled with petrol to Serbia; Greeks smuggled their goods to Macedonia; Serbians sold their goods to Macedonia while Macedonians sold their products on Serbian markets. Instead of fewer goods arriving in Macedonia, the market carried a plentiful supply of new goods, a strong contrast with socialist times. For my friends this situation meant, in particular, a connection to Europe through the trading of goods and also through the goods themselves. My female friends, by actually transforming their bodies into a representation of Western consumer culture, engaged in an intense discourse with the society surrounding them, particularly regarding the way this society connects to the outside world—implying a distinct change in the political meaning of economy.

CONSUMPTION AND TELEVISION

Much of my life in Skopje was spent watching American television programs including *Beverly Hills, 90210* and MTV (Music Television), but German and Italian advertisements were our first choices. We also watched the Italian version of *Beverly Hills, 90210*. If only as a diversion after work or before going out to the cafés, as a soporific or as merely background when visiting friends or relatives, the television was always on. A few people said that there was something strange about the place where I lived, and they always took a few minutes to work out what it was: the television was turned off. When television was discussed, people always mentioned how nicely those on TV dressed and how beautiful they looked, a factor much more important than the actual storyline, hence *Beverly Hills, 90210* in Italian. For a story, my friends go to the cinema or rent a movie, most of which are pirated. Television, however, is an important resource, a way of "knowing the world outside," as an informant put it. What is understood as "inside" is revised in terms of this knowledge of the outside world. This process can be termed "consumerism," a continual increase in the consumption of goods. The verb "to consume" means to eat or drink or to destroy, and to use up is *troshi*, which means to crumble, to spend, to waste; for consumer goods, one uses the phrase "*stoka za shipoka potroshuvaik*," which means "goods for broad consuming" or "goods for dropping money on." However, as Gell rightly argues, "consumption as a general phenomenon really has nothing to do with the destruction of goods and wealth, but with their reincorporation into the social system that produced them in some other guise" (Gell 1986: 112).

Consumerism in Macedonian has, in fact, much more in common with Pacific cargo cults, than with consumerism in the West. In this time of uncertainty and powerlessness, Macedonians receive messages through satellite television that there are powerful outsiders who have promised cargo goods. This cargo consists of Western consumer goods and images that are seen as metonymic of a whole system of power, prosperity, and status (Appadurai 1986: 52). My friends have internalised the idea that if Macedonia manages to establish itself as a modern European country, they will then "gain access to this cargo," which they feel will re-establish their dignity, and the social changes their country has gone through will then make sense to them. By making themselves worthy of Western European consumer goods, my friends seek to be worthy eventually of recognition as Europeans, by Europe, and to live as a part of Europe. However, although the imagery of the "cargo cult" portrays nicely the process of identity creation through the buying of "Western" consumer goods, in contrast to those described by Appadurai (1986: 54), my friends are aware of the economic forces behind the circulation of goods. Therefore, not a Macedonian "cargo cult" but consumerism identifies the desire to be something other than "former-Yugoslavian." This consumerism cannot be understood as copying another form of commodity exchange because, through their consumption, my friends are miraculously transferred to that land over the rainbow where the rich and beautiful live, that land free of economic misery. There is another world outside of Macedonia where one can become the person one wishes to be: beautiful, independent, popular, and respected. In the moment of consumption, then, in the moment of buying and admiring specific consumer goods, my young friends fleetingly inhabit this different world. As with the Muria described by Gell (1986: 136), material elements are selected and integrated into an immaterial cultural matrix, a collective style that becomes tightly integrated with the process of identity formation.

Consumerism in Macedonia is not to be equated with the "Americanisation" of Western Europe but arises, instead, from the very specific circumstances of an ever-changing Macedonia. Consumerism, the embracing of Western goods and ideas, existed before independence. However, today foreign goods and ideas have a specific meaning for a country that has declared independence under the distrustful gaze of Europe and in spite of the opposition of its closest neighbours. In Macedonia, to consume is not to express a longing for goods, access to which has formerly been denied; it is not an attempt to redress the balance between "twin" countries, as it is for East Germany. Consumption in Macedonia relates to the creation of a new understanding of what being Macedonian means. Macedonia does not want to be seen as a peasant society, as some of its neighbours portray it, and the attempt by some Macedonians to "return to our roots" was rejected by most of my friends. My friends want to see Skopje as a cosmopolitan society where objects of enchantment and desire are to be found in any Orka, Yukan, or Benetton store. Sorabji (1989: 45, 47) notes that, for the population of Sarajevo in the 1980s, especially amongst young, second-generation migrants, there is a feeling of superiority in comparison to those from rural areas. In addition, "The rules of Sarajevo are deemed by Sarajevans (whether native or migrant) to be different from, and better than, those of the country." Tone Bringa reports that in her study of Bosnian Muslim villagers, the city was seen as distinguished from the countryside through the absence of faith or religion (1995: 60). She relates this idea to the concept of culturedness, of not being a *seljak* from the countryside, that is pervasive throughout Bosnian identity and that describes the differences between the village and the city, as well as the differences between the villagers themselves. Nevertheless, the idea of this essential cosmopolitan identity was inherent in the Yugoslav system and is maturing today in a strengthened concept of consumerism. When, after Yugoslavia's disintegration, the cosmopolitan world (seen as the urbanised population's right) grew closer to the rural population of its own country and further away from the cosmopolitan world of Western Europe, the urban population of Skopje increased its consumption of Western goods in order to regain the idea of a cosmopolitan Skopje.

Nevertheless, Macedonia is not a carbon copy of Germany or the US in terms of consumerism, and my friends feel very strongly that their lives are a continuation of certain Yugoslavian ideas and values despite the influence of imagery transmitted through Western television series and advertisements. Considering self-construction, these images certainly influence the process of self-formation:

> … but as this process is located in the realm of images, reflective acts vis-à-vis goods are in a continuous relationship with other types of reflecting, especially other people and, of course, the whole realm of mass media representations, from television news and soap operas to movies, music videos and advertising. These experiential goods, whether distributed free or sold, may very well be consumed as "mere experiences" but they also act as mirrors for the aforementioned self-reflexive contemplation. (Miller 1997: 4)

Images and experiences drawn from Western television in Macedonia do not create or represent identity but are in themselves consumer items with a specific value attached to them, a value that can be very particular to Macedonia. An advertisement for Nivea hair shampoo might create a beautiful world of a happy family living in a single-family home and owning two cars and a dog, but this image is read by my Macedonian friends, as is any other Western consumer good, in a specifically Macedonian context. To buy Nivea shampoo is not to recreate the "Western World" but to buy a shampoo that is believed to be gentler for one's hair than the Macedonian brand, which is sold cheaply on the green market. Having well cared for hair is this advertisement's promise, but in Skopje this is understood in a Macedonian context. Consequently, Macedonia is creating a very particular image of itself in which the old is not cast out, but is given new meaning by objects that are more than mere luxury items and that, in fact, have come to represent the "Western" face of Macedonia.

In 1988, many of my friends, contrary to the prevailing attitudes of youth in Skopje, were sceptical of the idea that the country's young people were dedicated communists, hard working, enthusiastic, and socially involved. This political apathy was due to the feeling of not being able to influence social and political events. Positions of social, political, or economic importance were filled by people of the "partisan generation," and then nepotism ensured the transfer of power to their friends and family. My friends feel the same sense of disempowerment today, only the source of that disempowerment is different. They dissemble these feelings, however, through the purchase of Western luxuries, objects that seem to embody a feeling of power. In reaching towards these items, they reach towards the power that created them.

THE CHARACTERISTICS OF CONSUMERISM IN SKOPJE TODAY

Indifference distinguishes my friends from their own parents at the same stage of life. Partially, indifference is caused by their dependence on parents because of the altered economic situation of Macedonia in the nineties. At home, young people have minimal responsibility while parents try to give them a security they themselves did not experience. Parental expectations are simply that children be successful in school. The education system does not promote a sense of responsibility, and many young people in their mid-twenties see their main goal in life as partying. They expect pleasure from life without restrictions or responsibility. These attitudes, seemingly the direct opposite of those promoted by communist ideology, have endured and, although people's attitudes certainly change with age and starting a family, the social and economic isolation of independent

Macedonia and the sense of fear and insecurity this isolation brings encourage young people to live for the moment. Private aspirations, one's circle of friends, material possessions, and fashion are the foci of interest because of this feeling of disempowerment. This time however, it is not the state that is the source of feelings of disempowerment, but the fact that the state itself is disempowered. Macedonia exists because of the mercy and not the respect of Western Europe and the United Nations and because both fear that, if Macedonia ceases to exist, the region will be destabilised. Macedonians envision several scenarios that would destabilise the region:

1. Serbia swallowing up Macedonia thereby causing another civil war;
2. Macedonia becoming a part of Bulgaria thereby creating a dangerous instability between Bulgaria, Serbia, and Albania;
3. Macedonia becoming a part of Albania and civil war erupting;
4. Macedonia entering a state of civil war, with ethnic Albanians fighting ethnic Slav inhabitants;
5. Macedonia falling under Turkish influence, which would cause major problems with Greece.

As can be seen, Macedonia has a very sensitive position in the region, but this position certainly does not mean that it is master of its own destiny. Today's situation is markedly different from that of the time when Yugoslavia, under Tito, split from the Soviet Union and founded the Non-Alignment Movement. At this time, Macedonia within Yugoslavia was at its height in determining its destiny and the destiny of the region.

Consumerism in Skopje is not merely an imitation of something else, a copy of another political system. Material goods and images of these goods have become new symbols. This symbolism is strongly based on the negation of a peasant past, of a rural lifestyle, and of any similarity with today's Albanian-Macedonian population. Consumerism gives the young people of Skopje an idea of what they would like to see the future of Macedonia become. How the desire for this sort of identification is formed and what it implies demands consideration. What many of my friends choose from their understanding of modern consumption[12] is not a copy of a lifestyle they see on television or in Germany, but something far less coherent: a desire for freedom and proximity to a world that, for many, seems blocked off. Consumerism is their way of gaining, to some extent, control over political events. By bringing the West close to them through material possessions, they feel they are exerting a measure of control over their lives and are less separate from their Western European counterparts. Their patterns of consumption serve as a form of self-identification and should not be understood as choosing or copying a foreign lifestyle because my friends channel their consumption towards a specific goal—

gaining proximity to the European Community. They consume very specific products, images, or ideas and construct their lives around them. In the context of performances occurring in the Philippines, Cannell (1995: 254) points out that the apparent mimicry of Americanness is based on a different meaning of imitation in the Philippines than in the West:

> To take part in such a performance is both to move towards the pleasures of empowerment which come with "knowing the words" of a text and making it one's own, and also to move towards a transformation in which what is distant, powerful, and oppressive is brought closer and made more equal. In transforming yourself, in the process of becoming "beautiful," one also transforms the other.

Even though my friends seem to imitate "the West" through their consumption, their consumption creates something very specific and personal for them.

Not many years ago, we sat in cafés and drank Cokta, the Yugoslavian version of Coca-Cola, which was regarded as having a much better taste. Today, very few people drink Cokta even though it is one third the cost of Coca-Cola. Many cafés do not even stock Cokta anymore. These foreign consumer goods can be understood as the objects of an economic and cultural system that promises linear and circular times. Consumption becomes the outlet for the idea that Macedonia lacks its own life force because consumption not only represents the desire to be European but also constitutes its own very real social imagery in a way that does not always match reality. From this point, one can understand that self-definition in Skopje is contingent on consuming and consumption behaviour, on the idea that one's actual place in a global world is based on consumption. Mass media offers the way into places that are today physically out of reach. Gell's belief "that there is a valid distinction between dull, unimaginative consumerism, which only reiterates the class *habitus*, and adventurous consumerism like this, which struggles against the limits of the unknown world" (1986: 115) describes Macedonian consumption as well. Consumption is not the battleground of class distinction (and not even of straightforward ethnic distinction, although such a distinction is inherent in the concepts of "European" and "Other") but consumption serves as a struggle against limits set by social, economic, and political circumstances within the newly independent Republic of Macedonia.

The degree in which the notion of the "West" has become an instrument of creating "belonging" is remarkable. It is the world of "objects" and consumer goods that offers insight into this struggle to belong. The process of objectification creates a process of incorporation and rejection that defines "belonging."

NOTES

1. Most of these consumer "objects" had previously been brought in by guest workers or by people travelling. However, in the new Macedonia, these items gain a "new" meaning because access to the countries that produce these objects has become very difficult.
2. For example, teenagers are in some ways defined by the fact that they frequent different areas than their parents do.
3. Many of the images to which they aspire are from American television. However, the "ideal" world for them is Western Europe. America is often accused of social Darwinism and Americans seen as only interested in themselves.
4. Indeed, offering *slatko* is a custom found in most Mediterranean countries, such as Greece, Bulgaria, and Turkey.
5. No one in Skopje failed to mention the Turkish occupation and the oppression of Macedonia that resulted.
6. See Cowan 1990: 68ff discussing the sociability of drinking coffee.
7. Something similar is reported by Sheena Crawford in her dissertation "Person and Place in Kalavasos: Perspectives on Social Change in a Greek-Cypriot Village" (unpublished PhD dissertation, University of Cambridge, 1985).
8. This means literally no pay—workers come to a factory in order to keep their jobs in the case the factory starts working again, but during the time the factory makes no money, no one receives paycheques.
9. See Verdery (1991: 25–26) as cited by Humphrey (1995: 56).
10. Many informants in Macedonia believe that Europe views Macedonia as an impoverished republic on the brink of extinction; a republic to be pitied and not an equal.
11. Bourdieu (1984) demonstrates that value in the contemporary West is not just a question of demand, but I wish to emphasise the difference between the actual assigned value in Macedonia and that in Western Europe. Certainly one can find extremely expensive scarves in Parisian shops that only a privileged few can afford to buy. However, for most of the middle class, there are options to buy scarves that have a similar value as that of the scarves from an exclusive Parisian shop, but at a more demand-adjusted price.
12. "Modern" and "Western" are used interchangeably by my informants.

Chapter Six

Silhouette: The Sculpted Body

In trying to understand my friends' motivation to carry on during the difficult adjustment to a new country and to new political and economic circumstances, I noted one central theme: the symbolic meaning attached to their bodies. It was not politics but the design of their bodies that moved them strongly. I have come to believe that their efforts to change their bodies were directly linked to the changes their country was undergoing. Their position in society differs from the position their grandmothers and mothers held, and it differs also from the expectations they had for themselves when Macedonia was part of the former Yugoslavia. In my conversations with these women, they often described their lives as lacking personal autonomy, a term and a meaning that derived from the American and Western European television they viewed, which had become a part of daily family life. There were changes in their country that affected them directly but over which they felt they had no control. In juxtaposition, everyday they would see many beautiful women on television, and these women appeared immune from the normal female bodily processes. Now, my friends are certainly not so naive as to believe that such representations are real. However, they do accept the general idea that life in the West provides perfect ways of concealing bodily processes and that it offers ways to alter the body to fit the images on television. When I arrived in the field with a set of Always sanitary napkins, they provoked great interest, and I was asked if I could spare one or two. The shortage in sanitary towels had, however, ended, and what was really wanted was the experience of a Western and therefore superior method of hiding bodily processes. Soon, all sorts of Western pads were found in the duty-free shops around the city and in other import stores. In this chapter, I argue that the changes in the daily routine of my friends, changes that are directly linked to Western images of the slender and fit body, suggest a way of managing their bodies. It is through managing their bodies that the graduates I will introduce are making their inner intentions, capacities, and dispositions visible to themselves and others (Benson 1997: 123).

1988 IN SOCIALIST MACEDONIA

In 1988, my friends were entering university. After they had passed the university entrance examinations, a group of these friends and I went to Greece for a holiday. The days were hot, and, at night, they danced at the local disco to the songs of Madonna. Madonna represented what some of my friends wanted to be: "She does not care" was their most complimentary comment about her. Susan Bordo (1997: 268) depicts Madonna as a heroine who refuses to be constructed as the passive object of patriarchal desire. "*Nema gaile,*" roughly translated as "it doesn't matter" or "I don't care," was the phrase used to express this sentiment. This kind of sentiment made my friends feel very much at the centre of things. They felt they did not need to care about the old traditional ways of their grandparents or about the ideology of their parents. Yugoslavia supplied them with everything they needed, economically and professionally, and with the freedom to want more. Life was interesting and revolutionary. Many of my friends joined the prestigious engineering faculty in Skopje. Each knew she would eventually meet the man for her and, in the not too distant future, have a family. This family would not look like the families in which they had been brought up. Instead, my friends would have an open relationship with their partners, and their children would be their friends. They would have their own flats. Their friends would live nearby and would always be available for a quick chat. Leisure time would not be limited by housework because housework would be shared equally with partners, and there would be less of it anyway what with microwaves and frozen food and a cleaning lady once or twice a week. Also, they would have a career and challenges at work and travel abroad from time to time. The women talked of such matters a lot that summer and in the following two years, and it seemed just a matter of time before these dreams would become reality. These dreams were at hand, so they celebrated in the summer of 1988.

These women felt they were defining themselves: who they were and who they wanted to be. They were young women ready to face what life had in store for them, and they assumed it would be good. In those years, my friends had discussions with their parents or families in which they declared that they did not intend to marry. A woman might say she intended living with their female friends until she met the man of her life and decided to move in with him, married or not. In these discussions, women wilfully stood against what they saw as the past: the lives of their mothers and grandmothers. Their declarations prompted mild laughter from their kin. Nevertheless, it was a time of protest against what my friends then considered the dullness of their mothers' and grandmothers' lives.

Madonna, the heroine of MTV (Music Television) and youth culture, stood as a symbol of beauty and assertiveness for my friends. The world surrounding Madonna gave them social clues as to how a world could be.

By watching MTV, my friends experienced the world outside Macedonia as one big television transmission, and, as recipients of this transmission, they were part of this world. People from all over Europe called in to MTV with a specific music request, and many callers were from Yugoslavia. Through television, my friends, "us," "we," and "Europe" intersected. The word most commonly associated with the years 1988–1990 is "freedom" according to my informants. This period had its peak in the summer of 1990, when Ante Markovich, through drastic economic reform, tried to help Yugoslavia's economy. The *dinar* was bound to the *Deutschmark*, and, that summer, people in Macedonia had never felt closer to Germany. The next year, they were further away than ever, and my friends had difficulty identifying with Madonna and those like her. These symbols of the West became a subject of longing, something my friends felt they had once had but was now lost to them.

1990 IN SOCIALIST MACEDONIA

One morning at the faculty of engineering, while the hot air stayed outside and the cool, fresh, sunlit air circulated in the hallways, there was much activity: it was election day for the student union. The election boxes were positioned in the classrooms. People handed in their votes to a background of friendly chatting. There was a feeling of belonging. Exams were finished, the stress was gone, and the summer holidays were starting. However, at this moment, nobody seemed really to want to leave. The effect would not last long, but, at that moment, everything good in life was there. Surrounded by friends, people felt secure and in control.

Some of my friends ran and were successful in the elections for the student union, which gave them a sense of being wanted. They enjoyed the feeling of being voted into a job in which they had confidence they would do well. Ina became the foreign secretary, and she arranged contacts with other student unions both inside and outside Yugoslavia. Her duties included arranging a Student Olympic Games for engineering students. Suse took up the task of organising the exam schedule and negotiated with teachers on behalf of the student body. In all these activities in and around their faculty, my young women friends experienced a time of personal freedom.

Going into the summer of 1990, Macedonia experienced an atmosphere of exaltation. Shops carried goods that had not been seen for a long time, and many Western goods arrived, which brightened the previously bleak shop windows. New shops opened: pizza places, croissant shops. Yugoslavia was very close to being "European." In 1990, everyone kept telling me that soon Yugoslavia would join the EU. The iron curtain had fallen and removed the darkness from an expectant Macedonia. One often overheard and read in newspapers: a new millennium of European enterprise and peace lay ahead.

1995

Silhouette is a "Body Sculpture Studio." Prominent signs displaying the stylised black silhouette of a slim female body against a purple background guide you to this studio on the third floor of one of the apartment buildings that are reminiscent of the buildings at the epicentre of the 1963 earthquake. The street is dusty; downstairs people sit in a café or go shopping in one of the old supermarkets, which has only recently begun carrying Western yoghurt. The hallway is dirty; the elevator squeaks and is full of messages written in the Latin alphabet. There is no sign on the door, but the clientele show you the way. A loud doorbell announces your arrival. They let you wait. Eventually a beautiful, slim and young woman opens the door. You might have seen her at the university faculty or in the cafés in the evenings. She recognises you and smiles. The leather seats are inviting. The goddess of the temple is sitting behind her desk facing you and looking at you with her eyes half shut. She is sizing you up. How much do you have to lose? How could we tone those legs? Those arms? She gets some of her index cards out, makes some phone calls on a cordless phone. Glossy, shiny magazines from Germany, France, and Italy are laying out the ideal: the women pictured in them are very beautiful. They smile at you, measure you: you measure them. The owner is fifty, but looks older: she has lived a full life. A tower of false blond hair is piled on her head, with some loose strands of hair surrounding her fake, dull face. She wears tights and a tight T-shirt.

You are given a diet plan in which you are instructed to drink only water and some soup for the next ten days. You are measured with measuring tape strung loosely around your too ample waist and thighs. You will pay an average monthly salary for the ten-day program. This program includes exercise on eight automatic exercising machines, which force you to move your legs up and down and left and right. You sweat. The last bench is heaven; you can lie there for three minutes as the machine gently shakes you. The next day you experience a different kind of shaking: electricity is passed through your limbs and stomach. In the same room, there is a big mirror and a pair of scales. Every newcomer is weighed in this room and her weight announced to the girl who writes the index cards. Everybody lies there shaking together. At the same time, you are forced to look into the mirror, forced to face your own unwilling flesh. One treatment uses electrical current to cramp your buttocks so hard that for a few weeks you have a perfectly shaped backside. For ten days, you are beautiful. Then you are measured again, and this time the measuring tape is pulled as tight as possible, showing all the inches you have lost. But you could have lost more if only you had kept to the diet. Every day you were reminded by the lady at the front desk: "Stick to the diet." A sip of milk was your reward for the day. It was too hot to eat anyway. When it was all over, you were told to come back soon, so you could erase your sins. It is summertime, and the truth will be revealed at the beach: everyone will see....

These passages describe my own experience at the fitness studio when I accompanied my friend Ina. Recently ten new body studios opened in Skopje due to demand and the freedom brought by privatisation. Most studios are in private apartments or in institutions like the public swimming pool, and they are often combined with enterprises such as bodybuilding and aerobics. They have come to replace the semi-private cosmetic studios that offered body hair removal and facials in socialist Macedonia. The difference between these two types of studio is explained by a cosmetician, an enterprising woman who had the first private cosmetic studio in socialist Macedonia, perhaps in part due to her husband's position as an important politician in the former socialist government:

> Women come to me to be treated nicely and with care, even the waxing. I do this for them. They come to me to be spoiled. Those body studios make you work hard, and they punish you. They put you down all the time: they look down at you.

In Yugoslavia, the socialist system offered planned comfort and allowed for semi-private beauty studios. In today's Macedonia, however, not comfort but firmness and shape are sought in order for women to compete with the body images of the West. This different objective is one of the central differences in the world of beautification today and in socialist Macedonia. Yugoslavia offered comfort and self-confidence. Today's body cult in Macedonia offers more restraint than comfort.

1996 IN INDEPENDENT MACEDONIA

In 1996, knowing the right action is more difficult because social rules have changed. One is not in charge, but instead more dependent on others because the world has become smaller. Although the time has come that my friends anticipated in 1990, in the warm summer of Yugoslavia, their dreams have been altered drastically. Now they are deciding whether to stay in Macedonia or to leave. Many think of going, but very few actually leave. In all this, my friends express their feeling of being caught, and compare this to those of the past, when they felt free. Let me return to Madonna.

Since the summer of 1988, the idea of "the body" has changed its meaning for my friends, and, consequently, they treat their bodies differently. In a Hans Christian Andersen fairy tale, a mermaid trades her tail for feet in order to love her prince. In doing so, she not only loses her tail, but also suffers the pain of a thousand cutting knives in the process. Rather like the mermaid, my friends, specifically the young graduates, experience the pain of living and achieving their goals. Most of these recent graduates have boyfriends. Many of them have found jobs or are looking for one, and few have decided to continue their studies at university. They are thinking of marrying the boyfriends they have had for two or three years.

THE WORLD THEY LIVED IN—YESTERDAY AND TODAY

Who are these friends of mine? They were Yugoslavia's young generation. While studying engineering or economics, they took classes in Marxism and Atheism. They did not believe firmly in socialism but in capitalism and democracy instead. They saw the lifestyle they wanted in Germany and in America through television. Many skipped Marxism classes. Their experience of the Communist Party was restricted to the experience of being a "pioneer," a member of a socialist youth club that was similar to the Boy Scouts. In it, they sang songs of partisan heroes defeating the Germans.

People still sing these songs for entertainment on a Sunday walk to the top of Vodno, the mountain range overlooking Skopje. This walk, taken by many people in Skopje on the weekend, still corresponds with a socialist body culture. In Skopje, politics has not been a topic of discussion for most people, not in 1988 and not today. People still associate politics with corruption and with self-serving politicians. However, most of my informants, when they look back at Yugoslavia, feel that, despite its political system, its economic stress, and the resultant high prices in the eighties, they were able to buy most of the things offered in the shops. Of course, at times not even necessities could be found in the shops. Nevertheless, they could always buy consumer goods on the green market. While my friends had little money to buy these goods, some were given pocket money by their parents, and others had scholarships worth approximately DM 50 ($18) from which they could buy a set of pens or perfume. They remember Yugoslavia as not always easy, but generally improving, peaceful, and always offering a way to get by. Today the financial situation is a great deal worse, even though many more consumer goods are available, and my friends are earning their own salaries of about DM 300 ($100) per month.

Despite the worsening financial situation, in recent years a number of more expensive foreign shops have opened. In Yugoslavia, jeans were bought either when travelling in Greece or Germany, although designer copies could be bought at the bazaar for DM 15 ($5). One could buy clothes at the market, take them home, try them on, and, if they did not fit properly, return them. There was a friendly atmosphere in the market, and traders remembered their customers. Today, one can go to the many shops offering the authentic articles. The differences between the old way of buying and the new can be explained through the concepts of what is "ours" (Macedonian) and what is "new" or "modern." In "our" stores, one can buy a coat, for example, and agree with the sales woman that payment will be made in three monthly instalments without an interest rate being charged. However, many of the "new" shops do not allow payment by instalments. Therefore, only a specific clientele with a large disposable income and the ability to make cash payments can patronise these stores. With greater frequency, my friends are shown that they are not among this clientele, but still they dream of joining this world, and they make every effort to prevent the gap between themselves and "modern" Skopje from widening.

In 1988, we went to a shop that sold household goods because we wanted to buy shampoo that was behind the counter. In order to obtain such goods, we had to ask the saleswoman to write a note for us, which we would then take to the cash-point. There, we would pay for the shampoo and return to the saleswoman who would take the paper from us, wrap the shampoo in rough paper and then hand it to us. The first task, however, was to gain the attention of the saleswoman. She was smoking a cigarette. Once she had finished, we told her what we wanted. She listened, then began a second cigarette. Having finished her second cigarette, the saleswoman gave us the note, and we went to the cash point where three saleswomen were engaged in a heated discussion about one of their husband's passion for younger women. We waited patiently until it seemed we were a nuisance to the saleswomen, one of whom finally took our money. Returning to where we hoped our shampoo would await us, we found that the first saleswoman had disappeared. We waited until she came back. She wrapped our shampoo, and we left. In all this my friend had stayed very calm and waited for things to unfold.

Both types of enterprise, the friendly, private green market and the state-owned shops, coexisted in Yugoslavia. For my young informants, the poor service experienced during the shampoo purchase demonstrated the differences between socialist, state-owned operations and private enterprises. However, my informants' parents considered the green market a relic of the past, while for my friends' grandparents, the market meant an abundance of food. Consequently, the parents would buy in the state-owned shops, the grandparents would do most of their shopping at the green market, and my friends would frequent both places, but preferred shopping in Greece or Germany on their holidays. So in independent Macedonia, the duty-free shops and their successors, the new private "Western consumer" shops, became the site of shopping for many young people in Skopje, whereas their parents and grandparents mostly shunned these shops. For instance, a friend of mine "needed" to obtain an expensive perfume. She knew about the imitations offered cheaply on the green market but still she bought the perfume in a duty-free shop in Skopje. She bought it at the duty-free shop in order to affirm and confirm who she is each time someone sees that perfume bottle on her shelf; she did not buy it to deny the fact that this perfume was beyond her means or to pretend she was a person she was not. This perfume bottle closes the gap between her and the West.

Nevertheless, in independent Macedonia, the connotations of private and socialist enterprises are partially reversed and lead to a confusion of values. Today, when my informants think back to Yugoslavia, they see the socialist state as a caring one. The privatised world of Macedonia, however, appears to them as a cruel world where they must fend for themselves. In order to do this, they have to correspond to a picture that is outlined for them on television: firm, self-confident, and smart. Many of my friends fear this picture yet still want to embrace it.

To understand this conflict of values in its historical dimension, I will illustrate them through two examples: one from socialist Yugoslavia and one from independent Macedonia. Seemingly, the terms "private" and "Western" are identical, to a great extent, when considering enterprise. However, from a historical perspective, "private" stands in contrast to "socialist" in Yugoslavia, whereas in Macedonia today, "private" stands for "Western European." The difference might not be readily apparent, but it lies in the historical perspective. Today my friends are missing the peaceful and promising times of Yugoslavia. Then, although not embracing some aspects of socialism, they did not feel deprived, but rather rebellious. In independent Macedonia, however, they are trying to grasp the concepts that will allow them a Western lifestyle while attaining this lifestyle seems more distant and cruel each day.

In 1990, a friend and I went to the post office to post a letter but decided to return home in the afternoon because the sun was mercilessly hot. We rested for two hours in front of a fan before walking to the post office, which closed at seven. We arrived at our destination around ten past six and found no one in the post office save a post office clerk, and a most unwilling clerk at that. When we asked her to please accept the letter, we were told that the post office would close at seven. We told her that we knew this but that, as it was not yet seven, we would appreciate having our letter sent. At that time, my friend and I had a short discussion about democracy and the market-orientated society, and we both concluded that it was the duty of the clerk to accept the letter. In our discussion, we jumped from socialist social norms as described in the shampoo story to those exemplified by experience of the green market and of shopping in Germany and Greece. We concluded that times were changing and foresaw the not too distant future of Yugoslavia as a member of the EU. Because of this future, we decided to uphold the principle underlying such a change. After a short argument with the clerk, we left the post office triumphant. Returning home, we blurted out our upsets and eventual triumph to my friend's father. We were shocked, however, by his disapproval of our little victory. In fact, he told us that we should have gone to the post office in the afternoon. His response and his daughter's disapproval of it illustrate the generational conflict that can be seen as typical of the change in values within this society. Those values, however, in times to come, would be shaken again and point beyond a generational conflict to a completely new set of social values.

When the foundations of Yugoslav society were torn down, Macedonia and its citizens found themselves in a new struggle for social values. In the same year, my friends and I tried out our revolutionary expectations on a changing economy in a few more encounters. We argued at length in a newly opened private grocery store that we should be able to buy a single can of mushrooms and receive change without having to buy more cans. In the end, we forced the saleswoman, who would have preferred us leaving without any mushrooms, to go to a nearby bank to change our money. We achieved this only by asking for the manager. The manager came and actually agreed with us. However, our confidence in the eventuality of a changed economy with new social values was short-lived, and, by 1993, whatever changes had occurred had been reversed.

In another situation, my friends queued for an hour to visit a crowded café. They ordered drinks and were standing, squeezed against the wall, when their luck changed, and a table became free right next to them. What a feeling to finally sit down! This joy, however, was short-lived as a waitress came by ten minutes later and told my friends that the table was reserved for her friends, whom she had brought with her. My friends immediately got up as they knew many people in the café and did not want to cause any problems. Questioning them afterwards about that incident and others like it, I was told that things were just this way, and there was nothing to be done.

If both socialism and the political apathy that came with the fall of Yugoslavia rendered my friends powerless, what is the difference between the two systems? In some ways, the feelings of powerlessness that my friends expressed were not really a change. In 1990 and the years following people had felt that things would change out of necessity, not because of their actions. In 1995, people found themselves back where they had been or, perhaps, even more deeply dependant on the goodwill of the few people running the country, people they would term the "new" people or *biznis joveks*, business men. The country was significantly smaller, and, politically and socially, there was less space for individual autonomy and fewer possibilities for personal development. In Macedonia, the old times are now regarded with a certain kind of nostalgia. In many of my interviews, very different people told me, "Things were not right then either, but one nevertheless knew what to expect and how to cope with life."

The feeling of powerlessness was more intense by 1995. By then, many of my younger informants felt that they had to leave Macedonia in order to survive and live a decent life, a life not necessarily better than what they had lost in Yugoslavia, but at least equivalent. The parents of those in their twenties describe their own youth as a time of freshness, ambition, and an orientation towards the future. These descriptions have no hint of powerlessness and are without bitterness and disappointment. However, friction exists between this ideal picture of their past and their children's lives. The socialist program anticipated a specific type of young person: one interested in social issues, atheistic, and willing to fight for the common cause of socialism. This image had very little relevance for young people in the 1980s who felt they had little influence on events happening around them. These feelings of powerlessness were not caused by the disintegration of Yugoslavia alone, but were probably exacerbated by it. By losing the framework of Yugoslavia, my friends lost their framework of orientation, ideals, and social norms. They even lost things to rebel against. Young people today are dependant on their parents for much longer than their parents had been dependant on their families, and they are not obliged to do many duties at home, only to excel in their studies. Often my friends commented that they had never grown up and worried that they would never be independent. In fact, they saw marriage as the only possible way to become more responsible.

The summer days of 1988 were carefree for my friends who were first-year university students then; these feelings lasted until 1991, when war broke out and Macedonia declared its independence. Between 1988 and 1991, my friends in Skopje had nothing to worry about, had no responsibilities or restrictions, and expected only pleasure from life. This life was, of course, very different from that of their parents, who had moved from small villages to Skopje where there were great opportunities for ambitious young people in the 1960s and 1970s building an ideal socialist society. When the economy slowed down and parents failed to realise their own hopes and ambitions, they transferred their efforts to their children and tried to give them lives as carefree and as opportunity-filled as possible. Consequently, their children could centre their lives on their friends, material possessions, and consumer goods. How one looked became very important, and it became especially important to look "Western." Socialist ideology and partisan glory were confined to heroes of past childhood games. From this perspective, one could state that my friends follow a pattern of adolescent development similar to that of any young person moving from a state of dependency on parents to individuation and independence from others.[1] However, I argue that in independent Macedonia, this carefree life of the graduates changed and, even though certain values such as looking "Western" and beautiful have been retained, their meanings were ultimately changed through feelings of pain, loss of control, and the desperate need to alter the status quo.

CAFÉ LIFE, BEAUTY, AND MEANING

Whether it was a sense of hopelessness, a lack of responsibility or a lack of independence, things had changed by 1993. The Stara Carsija, the old city plaza, where thousands of young people had celebrated life into the wee small hours, is now deserted. Is it fashion that has made other cafés the choice of young people, or is it because Stara Carsija lies within the Albanian Quarter and is not seen as Macedonian? Social life in Skopje has definitely changed. Café life still dominates today, but other places away from Stara Carsija have become more prominent, such as cafés in the new shopping malls, in the Tragovski Centre, and in and close to the Park in the north part of the city. Only the alternative scene still frequents a few places at Stara Carsija, but these cafés are by no means crowded. Cafés are the places where one can be looked at, where you can see and be seen.

One evening in the café Van Gogh, near the Park, a large crowd of mostly "beautiful people" stood inside and outside the café, drinks in hand, trying unsuccessfully, because of the blaring music, to converse with people squeezed tightly next to them. Outside the café, macho guys were driving up and down the street in their nice cars, accompanied by beautiful slim girls, talking out of the car windows to their friends. The loud music caused complaints to the police, but they had already come and gone, likely with a nice bribe. Or perhaps the responding officers were just good friends of the owner of the café, a son of a "big" man or *biznis joveks* who had been given the café as a hobby. The police officers would have a couple of drinks, the music would be turned down for ten minutes, and, after they left, it was business as usual. The government had tried to enforce a law that would close cafés after 11 p.m. in residential areas, but this law was successful for only two weeks. After this time, some inhabitants resorted to throwing water or tomatoes from their balconies onto the crowd below, which only added to the atmosphere of hilarity.

We heard through an interesting information system that some of our girl friends were inside the café. With much discomfort, I squeezed my way through, trying to avoid being burned by cigarettes and feeling faint because of the stale air. Susana greeted me with a smile and a nod, and I received a hello from Beti and a nod from the other girls. I managed to order a Coke and stood with them in the middle of the crowd. Conversation was impossible, so I did what everybody else did, swayed a little to the blaring music, sipped my Coke, and watched the men, who were looking in the direction of a group of girls, eyes gliding up and down. They were typical of the girls in Van Gogh: tall, slim, tanned, longhaired owners of long, shapely legs tipped by high heels. These girls could make normal women feel very short, ugly, and fat. I did not follow this eye conversation any further as I had glimpsed some friends trying to squeeze out, and I had a desperate urge to squeeze out with them. Maybe I could go with them to another café, as crowded as this one perhaps, but at least alternative in nature.

My friends had actually left to visit the washroom, and I waited until they returned, so I could signal to them that I was leaving. When they returned, we elbowed our way out. Outside my friends stopped for one moment saying how stupid they felt, they knew they were not ugly, that they looked good, but still they were upset about the guys' eyes going up and down those beautiful slim legs that did not belong to them. They started blaming themselves for not having legs such as those these men clearly admired, for lacking the self-control necessary to be slim. Then they said that it was natural for men to like shapely legs. The discussion ended with several of my friends confessing that they desperately wanted legs like those at any cost.

I have described this evening at length, because it was like most evenings—not only for my girl friends, but I think for many of the girls in the cafés. Here I saw Hans Christian Andersen's mermaid cutting off her nice mermaid tail in order to walk on two feet for her prince. What had happened? It is the female body that has gained public importance in independent Macedonia, in the cafés, in the media, in most conversations I heard. This enchantment with the body stands in relationship to the changing world of my friends. The body became the site where the battles of loss of control, pain, and the desperate need to change the status quo were fought.

It was the winter of 1993: Macedonia was going through its most difficult time. Independence had been declared, and Greece had begun to boycott Macedonia; on public television, Serbia had declared that Macedonia was now South Serbia; elections were to be held, and the Albanian population in Macedonia called for them to be boycotted. Serbia and Croatia had their own "body politics" in which the female body became central to nationalistic discourses. (See Zarkov 1997.) Life in Macedonia was indeed unsettled.

THE BODY

On my return in the winter of 1993, I was first introduced to a *tresnje*, a "shaker," a person with a shaking machine—a new sort of private enterprise. The *tresnje* I visited with a friend had brought her "shaking machine" all the way from Australia, where she and her husband had stayed for a while, as her husband was a famous football star. Her name was Vera, and she told me that she was happy to be back in Skopje because nothing was better than home. Now, with her experience abroad, Vera started a small business "shaking" fat off people and assisting in the ongoing fight against cellulite. To my shame, I had not, until this time, ever heard of cellulite, but of course I was made to realise that my legs were in very bad shape.

My friends convinced me. In the flat adjacent to Vera's, which she had turned into a beauty parlour, Irina and I sat. We were wrapped in blankets with electrical currents, conducted by wet towels, running through our bodies, and we itched terribly. We talked with Vera throughout these sessions, and she told us about her life in Australia, how hard they had worked and how little time they had had for themselves. Meanwhile we became slim and beautiful. We flipped through the Western magazines Vera had in her parlour and imagined how it would feel to look like Linda Evangelista and what we could afford to buy. In 1993, prices were not all that different from those of 1988, shopping malls and beauty parlours excluded. Nevertheless, there was a difference, and this difference lay in the extent to and the ways in which my friends sought to alter their bodies.

Foucault suggests that the body is directly involved in a political field in which power relations have an immediate hold upon the body. They invest it, mark it, train it, torture it, force it to carry out tasks, to perform ceremonies, to emit signs (Foucault [1975] 1995: 25). I believe that my friends felt alienated after the break-up of Yugoslavia and that they acted out this alienation on their bodies.

Boddy (1989) describes Northern Sudanese women and their *zar* cult. In their *zar* ritual, the women fall into a trance in which they often enact cultural strangers. It could be suggested that these women, through their bodily experience, gain a greater understanding of their own society and their position within it. In some ways, their ritual represents a comment on their society and gives those women a distant look at their society, when everywhere else such distance is restricted.

Following these examples, one could argue that, in the broadest sense, the boundaries of Macedonian society have been altered, that the boundaries of the body reflect this alteration, and that, in Macedonia as in the *zar* cult Boddy refers to, one can find a problem of embodiment. Susan Bordo describes something similar in her analysis of anorexia nervosa and bulimia in Western societies. Bordo links these eating disorders to an idea of the physical body as alien, a not-self. In Augustine's formulation, the body is a cage that confines and limits, and, as such, it is the locus of all that threatens our attempts to control (Bordo 1993: 144). My friends in Skopje see their bodies as alien, as Bordo or Boddy describe, so experiences of hunger from dieting or of pain from exercising, manifest themselves as sensations that derive from the outside, invading their bodies (Bordo 1993: 146). The outside manifesting itself inside the body is very similar to the way spirits in the *zar* cult are experienced. Hunger and pain are not experienced as forces that originate from within their bodies, so, in order to master such "alienation," they have to learn to deny their bodies. They refuse to eat and are, therefore, creating their body-shapes, winning over nature, and, in the end, hoping for the satisfaction of material transcendence.

In her work, Douglas (1966) demonstrates how the body is represented in a particular culture by looking at how the body is read as text. Mauss (1979) discusses body techniques and Foucault ([1975] 1995) views the body as a passive object upon which external power is exercised. Though I am considering how my informants see and treat their bodies in a particular moment of time, I wish to stress that it is they themselves, not society, who are transferring messages that derive from their surrounding social world onto their bodies. Women's bodies, then, are not simply to be read as "texts of cultures," as passively reflecting the values of their society. The young women themselves draw upon these ideas, "make them body," because such action gives them the appropriate expression of a personal conflict (Benson 1997: 143). What drives young people in contemporary Macedonia to seek specific body images and leads them to a hard regime of body exercise and starvation remains a question.

BODY CONSUMED

Csordas argues in his essay "Embodiment as a Paradigm for Anthropology" that the body leads to collapse of the conventional distinction between subject and object and that this collapse allows us to investigate how cultural objects (including selves) are constituted or objectified in the ongoing indeterminacy and flux of adult cultural life (Csordas 1988: 40). His insight allows me to identify the bodies of my informants as being constituted in the ongoing daily life of social change. The consumption of images on television and in glossy magazines, then, leads to an understanding of the body that supersedes appearance, but promises a whole new lifestyle. Appearance plays a particularly important role for Skopje's consumers. Appearance is the external inscription of identity, and this inscription is radically different for my friends than it was for their grandparents or even their parents. In the world of their grandparents, appearance was quite standardised, dictated by locality, folklore, and economic circumstances. In the world of their parents, appearance became part of an ideology. An informant pointed out that in the 1950s and 1960s, many more women wore trousers than do today (1988). In Macedonia in 1994, several sectors of industry required, in response to public demand, their female employees to wear skirts. When asked to describe the fashion of today, informants identified the way that they and most of their educated urban friends dressed as "smart." When asked to describe the way their parents and grandparents dressed, my informants said that their parents' dress was functional or practical and their grandparents' dress was "old-fashioned." These differing descriptions reflect far more than simple changes in fashion or ideology. Dress and other bodily adornments, make-up included, have probably always served, to a certain extent, to individualise one's person, as well as to function as a social statement of inclusion or exclusion. Not long ago, every village and community in Macedonia had its own specific dress and within this system of dress, married women and men dressed differently from unmarried women and men. Today, there is less organisation of dress and behaviour. This lack of organisation is characterised as "Western" by my informants. In this situation, my friends have to undergo a self-initiation in order to create themselves as citizens of new Macedonia, in order to transform their individual bodies into socially acceptable bodies. They do so by constructing their bodies in the image of the world they find in American television and Western European advertisements. The television is an instrument that helps them form their modern life, and, undeniably, the television they watch is itself produced to "create" the body and its desires. Advertisements for Always not only show my Macedonian friends which sanitary pad to use, but also present them with an ideal: a fresh, clean, lean body in total control of its bodily functions. It is a body worthy of a pair of expensive Levi's jeans.

A friend of mine, after extensive dieting, took me shopping. She described it as shopping for her new self because now she could finally look like the singers of MTV if she chose to dress like them. Our first trip was to the Levi jeans store on the high street. It was not the buying of the jeans alone that created her new body, but the many turns in front of the public mirror outside the changing room and the repeated questions to her mother, the saleswoman, and me about how she looked. It was not important that we thought her gorgeous in any form or that the saleswoman probably did not care; instead, the self-assurance created by the public mirror miraculously created a new self for her. She had proved her power over her body. Furthermore, she had started a continuing process of reflexivity. As Giddens states, "The continual reflexive incorporation of knowledge provides precisely a basic impetus to the changes which sweep through personal, as well as global, contexts of action" (1995: 29). Accordingly, my friend's reflexive response to her own image started to change her perceptions of herself and of the world around her. Furthermore, her changing perceptions called for differing social actions as well. The mirror had taken on a new importance in her life; it was this mirror image she could compare to the images on television. Her ritual should not be misunderstood as self-adoring. Rather, her control and her reflexivity created her self in the midst of a plurality of choices, choices that ranged from a life that her grandmother had lived to the world of MTV. With control over her body, she was able to start taking charge of her life. She had given herself the form she had sought by assigning herself a very specific lifestyle.

When asked to define this lifestyle, women said that the body would be shaped to withstand being "disfigured." The bodies of my friends' mothers and grandmothers were understood to have been "formed through life." My friends felt that their grandmothers had aged too quickly because of the harshness of their village lives and that their mothers were, through neglect and carelessness, bloated from childbirth. They thought that their mothers and grandmothers had not behaved responsibly towards their own bodies. When faced with such accusations, the older women would laugh at the foolishness of their granddaughters, while the mothers took pride in their daughters' bodies. Their energy had been directed towards different goals, labour and childcare, and they were happy for their young daughters to be given the chance of a period in their lives when they could care for themselves and their appearance.

None of the older women I talked to were opposed to young women marrying later in life, as they now choose to do, nor to the frequent change of partners, although premarital sex was still completely unacceptable. In all this, the question remains as to why extreme weight loss was considered helpful in achieving selfhood. The mirror was put next to the images seen on television and in Western magazines. The use of the word image is sometimes taken lightly, and I do not wish to make a connection between "images" in magazines and television and the obsession of the young people in Skopje with thinness. But cultural images themselves are powerful, and the way in which they become imbued and animated with such power is hardly mysterious (Bordo 1997: 113). Images are not "just pictures"; they have a great influence on the lives of my group of graduates. I argue that, with the disintegration of Yugoslavia, the susceptibility of my friends to such cultural imagery changed, and the perfected images in "Western" television advertisements and those represented by supermodels have become a dominant reality for them, setting standards that are unrealistic.

Interestingly, none of the parents were much concerned about my friends' obsession with thinness; it was the grandmothers who were worried about the unhealthiness of their granddaughters' lifestyles. I suspect that, in some ways, the parents themselves fall into the trap of wishing their daughters to be similar to the images they see about Western Europe. Bordo suggests that such a phenomenon is related to an increasingly image-dominated culture in which the presence of counter-cultural body ideals has become diluted (Bordo 1997: 118). Most of my friends' grandparents do not watch television, and live in the countryside and not in Skopje. The world of Skopje, however, has become clearly dominated by the significantly increasing "Western European imagery" since the disintegration of Yugoslavia. This imagery is not seen as reality, not even as the reality of "Western Europe." My friends are too "image-experienced" to believe this. Nevertheless, these images create a specific desire and a fantasy of wanting to participate in the world those images portray.

I had several interviews and conversations with my friends about their excessive efforts to adhere to specific cultural images. These interviews took place from 1994 to 1996, when my friends had graduated and were working and earning their own money, a large portion of which was spent on enhancing their bodies or on presents for friends, on entertainment, or on siblings who did not earn their own money yet. Only occasionally would they contribute a portion of their income to offset the family's household expenses, such as petrol for the car; speciality foods they enjoyed, such as Cornflakes or Muesli; or body care products for their own use. Their parents paid most of the living expenses, such as maintenance of the flat, hydro, and food. Earning their own money coincided with startlingly obvious changes in their bodies. They all started a diet, bought clothes in the "new" shops, and worked on shaping their bodies in one form or another. Names have been changed to protect their identity.

INTERVIEWEES

Suse, a 25-year-old mechanical engineer, lives with her parents and sister, works at a radio station, and started swimming and dieting eighteen months ago. She has lost approximately ten kilos. She loves dressing smartly and is admired by her male work colleagues for it. She has been involved in a serious romantic relationship for two years.

Eleonora, twenty-five, lives with her brother at her aunt's home because her family lives in the eastern part of Macedonia. Eleonora's parents send money to the aunt for the care of Eleonora and her brother. Eleonora has known her partner, Rade, for five years, and they are planning to be married soon. She works in a small private computer company. She exercises regularly and, from time to time, she visits body sculpture studios; Rade pays for these visits. She also likes aerobics. She is very concerned about what she eats and despises traditional Macedonian food. She has many arguments with her aunt, who cooks for her, about food. She does not want to tell me how much weight she has lost over the last year, but she has become very thin, even her fingers are very thin.

Maja, twenty-four, has just finished her last exam. Her father is looking for the right connection to secure his daughter a job in one of the ministries. She has no boyfriend, but fancies several men. She lives with her parents and her younger sister in the newly built apartment buildings of Aerodrom, a district in Skopje. She started dieting only a few months ago. She also went to a body-shaping institute, which was a graduation present from her aunt.

Irina, twenty-six, works in one of the Macedonian ministries and lives with her parents and her older brother in the centre of Skopje. Her older brother has recently gotten married, and his wife is now living with them. They are all getting along well. Irina has been dating her boyfriend for seven years. She jogs nearly every day. Over the last year, she has lost about eight kilos.

Eli, twenty-three, lives with her younger brother and her parents in the south of the city. She has been dieting for two years. She has studied economics and is now working at a private import-export company. She does not have a boyfriend. She goes to a bodybuilding studio at least twice a week, but tries to go every second day. She has also started a strict diet program imported from America and follows an aerobic program on television if she finds the time. She did not tell me how much weight she had lost.

Blagica, twenty-five, lives with her parents, her younger sister, and her grandmother in a small house close to the centre of Skopje. She has been dieting since she left university two years ago. She works at the post office as a computer specialist and would love to go to Germany for a computer fair. She has no boyfriend, but started recently to date a colleague of hers. She does some bodybuilding and goes to the "shaker" periodically.

Milka, twenty-five, lives with her parents and older brother in the centre of Skopje. She loves exercising and tries to go every day. She has just found a new boyfriend. She works with the police. She is very slim, loves to go to aerobics and to the Olympic pool to swim.

Valentina, twenty-six, lives with her father and older sister in Skopje. She works as a technician in one of the ministries. She has no boyfriend and is very frustrated about her weight. She started doing aerobics only several weeks ago and has visited a body-sculpting studio once.

Susana, twenty-five, lives with her parents and older sister in the outskirts of Skopje. She works in a private computer company. She has been dating Vlatko for two years. She started dieting two and a half years ago. She has lost about ten kilos. She regularly goes to body sculpturing, bodybuilding, and to a "shaker." She runs and swims.

Elizabeta, twenty-three, lives with her parents and younger brother. She is in her last year of computer engineering. She has lost 3 kilos over the last two months.

Aneta, twenty-four, lives with her parents in the Aerodrom district. She finished her studies recently and works in a company owned by her uncle. She has been dieting and going to aerobics for a year now. She has a boyfriend whom she has known since high school.

Sandra, twenty-five, lives with her parents and two sisters in the centre of Skopje. She works as a computer specialist in one of the ministries. She has been involved with Filip for five years. She started to watch what she eats and to exercise two years ago. She goes to body sculpturing and bodybuilding.

Mirjana, twenty-six, lives with her parent and younger sister. She plans to marry Mitko soon. She works as a mechanical engineer in one of the ministries. She goes to bodybuilding, sometimes to a body sculpture studio, and she swims. She has lost 11 kilos in eighteen months.

Despina, twenty-three, lives with her mother and aunt while her father and older sister live in Belgrade. She has been with Stevo, who is twenty-five, for four years now. They are planning to get married soon. Stevo got a job offer that would send him to Amsterdam for a year. She started dieting a few months ago. She goes to a body sculpture studio and to a "shaker" regularly, has started to do aerobics, and runs after work. She works in an import-export company.

In discussions with my friends about their dieting and body shaping, they tried to explain some things to me:

> Suse: "I have always been frustrated about my looks. I think I can change this now, because of Borche [her boyfriend]."

For Suse, hunger and sexuality have become confused. The message she is reading is that a woman's body should disguise the parts that make it distinctly female, its particular physiology. Slowly her experience of her body is changing. Through Borche, she learns that a slender body is desirable and that with her more androgynous form, she can enter a man's world, a world that demonstrates its strength through control. Suse and her friends, through control of their bodies, demonstrate such control. They see the female body as especially difficult to control. The experience of their bodies and of their possibilities sheds new light on their social world. Being in control, being slender, secures them a place and recognition in a man's world.

> Eleonora: "You see, people realise that I have lost weight. My boyfriend does not think I should lose weight as he thinks I am very skinny. It is only my legs that trouble me. Right now I am trying to eat healthily and dieting for a while. I do not know how long I will be able to take that because it is hard sometimes; it really bothers me: I feel hungry sometimes, and it is not a nice feeling. I think exercising is what I need most. But I need the dieting to lose a bit; there is fat, which goes away. I know Rade likes girls with skinny legs. I have never liked my legs."

Eleonora talks about possibilities and strength and, like Suse, she feels that the world around her is changing and that this change means she has to change too. More importantly, in this new world, she suddenly has the means to change her body. She has a supportive boyfriend. She has the shops in which to find the clothes she sees in the glossy magazines. She earns some money and spends it on herself and her body sculpturing. Classes in aerobics and bodybuilding are offered; these have only recently appeared in Skopje. She knows the images of the Western world that tell her "If you become like us then you belong to us." In a world that has redefined her country's physical boundaries and, in the process, greatly limited them, this last message is an important one. Her body becomes her enemy because it is different from the images in the glossy magazines. It is her body that confines her to a world where she does not want to belong.

> Maja: "I am a person who likes eating. There are some people who do not care about food very much. That's why I try to arrange not to be hungry, but still if I would eat more it would turn into fat. I think right now I do this to prove something: that I actually can lose weight."

For Maja, the answers to questions about her life are found through control. If she can control her body, she can control the world around her. For her, control is not just about power, but also about her ability to suppress her hunger and desire. A slender body, the achievement of her final goal, will be the triumph of will over the physical. Her body will be pure and ruled by her mind. She loves to see herself as a scientist, and she would like to go to international conferences on computer engineering. She hopes to be a rocket engineer. For Maja, the rule of the mind includes the denial of her sexuality.

> Irina: "I see the fat. I see fat on my thighs. I measure myself, and, if there is one centimetre less of fat, I will be fine. Then I do more exercise and eat healthier food. Also now at least I eat: I eat fruits, I eat a banana, I am not starving."

Irina has a mother who loves her dearly but expects a lot from her in her studies as well as in her physical appearance. She makes a lot of choices for Irina. Irina's physical transformation was greeted with respect. Irina started to feel more confident with her "new look." She started to buy expensive clothes that emphasised her shape. She felt that she had subscribed to a completely new lifestyle, a lifestyle in which her mother had no part. She hoped to leave Macedonia for America or England where she felt she would find fulfilment. In contrast, she saw her mother as unable to change her own life.

> Eli: "I try to avoid unnecessary calories. I think once you have achieved a healthy look, you can keep it. But you keep it by being more physically active, by exercising. You always have to be careful what you eat. Of course, if you eat cakes every day, you get fat again. You just have a piece of cake twice a week. My mom says I am ruining my health, but I am eating salads and drink yoghurt and all these things. I do not have to stay hungry and I am exercising my body."

Eli's world very much revolves around avoiding the unnecessary, avoiding being careless. She has a sense that there is danger around her, which she confronts by exercising control over herself. She feels that, if she maintains this control over herself, there will be no part of her vulnerable to attack. Her body has become her sole world; all her thoughts circle around calories and how to reduce her intake. To know how far she can go gives her a feeling of satisfaction.

Eli and her friends differ from many anorexics in that they seek not to destroy their bodies and to become invisible, but to exercise control over their bodies. This motive is in contrast to what Orbach (1978) identifies as the central reason for women's ill-treatment of their bodies through their attitudes toward food; that is, women's lack of power and fear of violence and harassment. What I found with the people I interviewed is that many of them not only diet, but also exercise, run and go to aerobics and body sculpture or bodybuilding classes. They do not hate the female body, but instead try to control its image.

> Blagica: "In the morning I eat a banana and an orange,[2] and I am still hungry. But I think it is enough food for me; it is only that my stomach is used to getting more food maybe."

Blagica's self-denial is typical. However, her grandmother, Milka, disagrees vehemently with her ideas about dieting and often argues with Blagica and, indeed, tries to entice her to eat. For Milka, starvation does not mean control, it means to be at the mercy of something else, the elements, nature, or an occupying force. Blagica knows the feelings of powerlessness too. In order to apply for a visa to visit Germany, which was eventually denied, she had to wait in line to beg to be let in to visit, and, not surprisingly, she felt degraded. Blagica insists that her body, the way she shapes it, proves that she is different from her grandmother and the world her grandmother knows. Blagica has always insisted that she would not end up like her mother doing all the house work, being submissive to her husband, and controlling of her children. By sculpting her body, Blagica voices her rebellion against the ultimate fate described by her mother and grandmother: passivity and acquiescence to the world around them. She has a force driving her, sometimes she can feel this force in the form of hunger, but, more importantly, this force is visible to her when she looks in the mirror.

> Milka: "I exercise. I go to aerobics now. I swim. But what I wanted to do is to run a bit more, not just once a week, but I need company for that, so I will see how that works out. I can also exercise at home, but I become bored with that. I do not like it."

Like Eli, Milka is driven by her desire to strengthen her body, to not allow her body to be weak. She is less obsessed with leanness than with actually feeling good about herself. She loves shopping and to buy nice things for herself. However, she feels she needs to deserve these treats; this is why she exercises. She sets herself goals, and, when she reaches them, she rewards herself. These rewards are always in the form of an enhancement of her effort: she will not buy chocolate for herself, but rather a pair of tight jeans that now fit. She is cheerful, refusing to take anything around her too seriously, especially politics, and she is very popular among the young men of Skopje. In their eyes, she is the perfect woman: confident, amusing, and someone who achieves her aims with seeming effortlessness, as I was told by some of my male informants in different conversations. She is the embodiment of the images on television promising worry-free freedom, Macedonian-style. Valentina presents herself very differently from Eli:

> "I want to go to Ohrid for a weekend maybe, but I would not like to go there for a holiday. In Ohrid, everybody looks at you; I am tired of that. I just do not want to see anybody from Skopje. I do not want people to see me in my swimsuit. I would rather go to Cyprus where no one knows me."

For Valentina the body and freedom to travel are related. Indeed, she feels that the most popular conversation in Skopje, where one will spend the summer, is directly related to a specific body culture. She feels that "the summer body culture" is crueller than the "coffee house culture," which at least allows one to disguise one's body to a certain degree. On the beach, the swimsuit reveals all; the body is exposed, open to the eyes and the judgement of all. She associates freedom directly with places outside of Macedonia, with being away from Macedonia and free of the restraints and expectations she feels are imposed on her body there. She felt she was, for the first time, a woman in her own right when, in 1995, she left Macedonia and visited Amsterdam.

> Susana: "I know he loves me, but I am scared that, maybe, he will see someone he likes more and decide he does not like me anymore. Sometimes my legs might make him unhappy, but I think, if I lose weight I will be completely confident about myself. Then he has to be scared that I may see someone. I can imagine myself, when I come to this point, it is scary. Before I would forget about my problem with how I look, but then I see the girls with short skirts, which I really like, and it makes me upset. It is really making me upset. I want to be able to wear shorts and not always be conscious about how I look. And now my weight bothers me even more because I think maybe he sees these girls and likes them more than he does me."

Susana spends a lot of time worrying about her legs, neglecting her career, her family, and friends. She is willing to sacrifice everything for a beautiful pair of legs and the recognition from men that this brings. So she diets, exercises, and sculpts her body to achieve the final goal of shaping her body perfectly. In seeking love, she accepts that it is legitimate for her boyfriend to have a problem with her legs because she feels it is her fault that her legs are not thin enough—she has not shown sufficient self-restraint. Criticising herself, she often refers to herself as wanting too much. Susana's mother emphasises this every day: "She wants too much. When will she learn she cannot have it all?" So Susana does what is expected of her and restrains herself.

> Elizabeta: "It is difficult to find a boyfriend if you do not look a certain way. It is something you cannot change."

As with Susana, Elizabeta is very aware of the social implications of her own body. "How you look is what you are" is the common assumption in the café houses of Skopje. Elizabeta voices the desire to embody a specific image of a woman, a dominant image in her society and one that women themselves have a central role in creating. A common saying in Skopje is that men only want to have sex, women only want money. Turning this around, women represent the sex object with its touch of Western imagery, whereas men supply the financial means for such imagery, the imagery my friends see on television in Skopje, in American soaps and German advertising. There are no victims in this game, no losers and no winners.

> Aneta: "*Beverly Hills* is a colourful series. I like watching it, and the people look nice. It is relaxing; you do not have to think. I have always been obsessed with clothes, and they have interesting ideas, very nice. Their tastes in clothes are fun. My mom does not like how they dress."

Many young women in Skopje love to watch *Beverly Hills, 90210*. It is such a happy world, light and easy to digest. Women look beautiful and men are rich and drive fast cars. Nothing is serious: no serious work, no serious effort. The show does, however, have a serious impact on the people in Skopje who watch it. It is no accident that the new shopping mall is called Beverly Hills. The images presented by the actresses tell my friends how to hold their bodies, how to shape their bodies, how to behave, and ultimately how to express themselves through their bodies. By shaping their bodies in a specific way, my friends express what they cannot express in words—their reaction to the demands made on them and the feeling that their world is in turmoil.

> Sandra: "I was talking yesterday with my colleague, Bojan, at work, and he said, 'Come in a short skirt, so we can judge if you are fat or not.' He said this in a nice way. He says, 'Come on, I am your colleague. You do not have to be afraid. I just want to know how you look in a short skirt.' He is conservative. He thinks girls should wear short skirts. Right now I do not feel comfortable wearing a skirt. I would wear a long skirt, maybe."

Sandra feels that men make the demands on women's bodies. Men like feminine women in short skirts. By tailoring her body to fit the image of the ideal woman, she is able to gain entrance to a man's world. She experiences the empowerment that comes with a sculpted body, a body image representing the free, powerful, and worthy world—a world not even the men around her can easily access.

> Mirjana: "Mitko does not look perfect, but I am fine with the way he looks. My body does not bother him when he is with me, he says, but maybe when he sees someone else, it does. I can see him looking at other women. So he is helping me with my exercising. He says he does not want me to do it for him, or for me, but for us. In the beginning of our relationship I thought my body disgusted him. Right now we are fine; he likes everything else on me. I have never had anybody who cared so much for me."

Mitko and Mirjana's relationship mirrors a new set of values in Macedonia. The body is seen as demonstrating correct or incorrect attitudes. When I spoke with Mitko and Mirjana, it became clear that Mirjana's excessive exercising meant something different to each of them. Whereas for Mitko, it is primarily about the containment of Mirjana's female qualities and weaknesses, for Mirjana, exercising is a form of escape from the kind of confinement her mother experiences. Her mother's body shape associates her with a domestic and reproductive destiny, a destiny Mirjana is not willing to share, at least not yet. She "wants to be out there and be seen."

> Despina: "I am doing something I always wanted to do and did not previously have the will to do. If I achieve looking beautiful, I would be completely happy. If Stevo then decided to leave me, it would be too bad for him. I would be very confident about myself. If you think rationally, you can do anything. Only I think about food all the time."

Despina and her friends have reacted against the world of their mothers, a world they resent and that, for them, fails to correspond to the world of "young, free, intelligent European women."

In their specific, but also very similar, responses my friends show the importance of their female bodies in Macedonian society today. I suggest that the cultural discourse my friends are involved in is about what their desired body evokes. Their bodily appearances convey a specific meaning, a meaning that has been changed by recent political occurrences. The control of weight and the choosing of a "healthier" lifestyle have come to stand as the opposite of traditional, Macedonian values, values that have regained importance in independent Macedonia. In contrast to being beyond desire (Bordo 1997: 127) and in contrast to Bordo's earlier assertion that slenderness derives from the wish to be less female (Bordo 1993: 148), my young informants in Skopje seek to be to influence the images of being seen as modern, European (non-Albanian, non-Balkan, post-socialist) Macedonian women. They have very concrete desires, desires that, to a great extent, point beyond the boundary of Macedonia. Their hunger, though they triumph over this need, symbolises a very strong desire. It is not coincidental that a rumour went around that the New Zealand Immigration Department was accepting only intelligent, beautiful people, preferably engineers. Access was directly related to beauty. It was said that the agencies negotiating with the government of New Zealand were asking for thirty-nine pictures of each applicant in order to process the application.

The young women I spoke with were not so much concerned about the sexual aspect of devouring, as they did not feel they needed to control their desires, but they wished to control the world around them in a time when they felt they had lost control. In particular, they had lost the control to rebel against the world of their parents. They could not go out and think of living an independent life in Slovenia or Serbia or perhaps even in England as successful engineers. They were not in touch, either, with the material means of such independence or with the world they had before termed "ours," the world of young, striving Europeans. Certainly many times this idea of "Western" or "other than Macedonian" showed itself as an illusion. The sister of an informant won an American green card through a lottery in Macedonia.[3] She went to America, only to come back disappointed. She told people in Macedonia that Americans dress badly, work all the time, and never go out to cafés, and that's why she had to come back. Another person, who had gone to Toronto, reported back home that he just could not believe how badly the Canadian people dressed, indeed how they dared to go out on the streets wearing what they did, and that they did not know how to live life.

Reality is not always in accordance with the images people have created of it. But, in creating this idea of "Western," my friends create something very distinctly Macedonian: the Macedonian idea of what "Westernness" contains. Adrienne Rich considers body shaping an assimilation: "Change your name, your accent, your nose; straighten or dye your hair; stay in the closet; pretend the Pilgrims were your fathers; become baptised as a Christian; wear dangerously high heels, and starve yourself to look young, thin, and feminine; don't gesture with your hands ..." (1986: 142). However, assimilation does not describe the world of my young starving friends. Images of Western Europe and North America are not used in order to assimilate to such pictures; "thinness" and beauty have very different meanings in Skopje than in North America. The image in Macedonia of American or European life is a creation of media and imagination, which is why many people are disappointed with what they see when they experience a country outside Macedonia. In Skopje, beauty and thinness do not only refer to the external but are also an inward reflection. Even though slenderness might come to stand for a successful assimilation of a "Western" lifestyle, this lifestyle is created not only to diminish the difference between Macedonia and the "Western world" but also to create a difference between my young informants and their parents. The difference is created by an outward assimilation to a world that is today entirely inaccessible[4] to the parents' generation whereas some of my friends still feel that they may be able to realize their dreams. Through their bodies, my friends try to establish an image of a seemingly "free" woman, modelled after those they see on television, ready to claim her space in society and very different from her mother. In a time when Macedonia as a country is establishing its roots, they are adamantly opposed to falling into the roles they have seen their mothers occupy, roles defined by a peasant origin, according to my friends.

Essentially, the transformation of the body is an ongoing attempt to master their world. For some, it becomes compulsive and extreme; how far it goes depends on the individual's personality. For them, their exercising, aerobics, running, clothes, make-up, shopping, and body sculpting bring a sense of achievement, which, in turn, engenders a sense of empowerment. The source of this empowerment lies in the management of their lives, which they see as achievable through the transformation of their bodies. It is important to note that my friends are combining their dieting with aerobics, body sculpting, or bodybuilding. These forms of transforming the body are linked through the extreme and excessive enactment of Western culture's fantasies of the "self-controlled body." Ultimately, my informants redefine the relationship between what they believe to be truly them and a representation of themselves in their social world, which is constructed through interactions with others and the wider cultural framework through their body acts (Benson 1997: 125). As well, my friends are concerned with the issue of health and healthy eating; an issue they feel is not addressed in their society. Through television, they come in contact with the "modern" issue of healthy eating, health food, organic food, and herbal remedies. By following this popular discourse, they redesign themselves by having the same concerns as their peers in Western Europe. Their image of the "healthy self" is sustained, in part, through the creation of the image of their "unhealthy" mothers.

WOMEN BEFORE AND UNDER SOCIALISM

Discussing the bodies of young women in present day Skopje raises the question of how their mothers' and grandmothers' bodies were formed. To understand the difference between the younger women's generation and their mothers' generation, one can compare the duty-free shop and its Christian Dior lipsticks with the drugstore and its homemade cucumber-camomile cream masks. Honey and beer are good for the hair as they make it strong and shiny. Lemon juice is good for thin hair. Vinegar makes the hair soft. Cucumber or egg white makes your skin feel soft. One recognises the age of a person from the shade of lipstick or eye shadow she is wearing: "There is only one kind of lipstick, eye shadow, or hair dye, and everyone is using it" (Diana in 1992). The pharmacists make perfume, and, as long it is perfume, it does not matter how it smells. One day Eli and I went to buy Paloma Picasso perfume for Mothers' Day. For Eli's mother's birthday, we gave her a skincare program by Estée Lauder, products which we bought from the duty-free shop, a gift from daughter to mother, which ended up in the daughter's closet. The mother felt more comfortable following the old beauty recipes that had been given to her by her mother and her grandmother before her, possibly on their wedding days. But several women her age did welcome today's changes. One middle-aged lady described how she saw her life in Yugoslavia: "Tito gave us freedom without beauty. There was only grey." Women of her age, the same informant conveyed, used "natural" enhancers for their wellbeing. As their country sacrificed luxury for freedom, women sacrificed their comfort too. This sacrifice included not only make-up, but also simple sanitary products, such as sanitary pads or even toilet paper. "I will never forget where my family came from," another middle-aged lady told me.

Their world was that of peasants: no electricity, no running water, no beds, and no education. That they had left this world behind in search of a better life and that this better life could not be found without sacrifice was the commonly understood ideology.

Socialism had brought them so much; it had taken away the hardship of peasant life and foreign occupation. In any discussion, of politics or beautification, this was always mentioned:

> We learned to be happy with what we had and to be proud of what we had achieved. We liked to worry about small things: how to make a nice home or dress nicely. But we knew that in comparison to those of the whole country, these were small concerns. My children today worry about small things too. They like to go on holiday; my girls like to look nice. I help them as much as I can. There are so many terrible things happening. What else can you do but make your children happy with small things?

Zorica believed she had controlled her world in the past; people then were concerned with the building and formation of society. The world today, she describes as being out of control, and she cannot name the people who have control. When asked, she would say that America and Germany first plotted and then caused the break-up of Yugoslavia, and finally took control. In helping her children to obtain small gadgets from the world, which controls them, she hopes to empower them. Looking back at her young adult life, Zorica says she liked to look nice but was told to use her energy in the pursuit of different goals. Today, for Zorica's children, embodying the image of control is important, and her daughters have to control what they can: their bodies. According to Benson, Bynum argues that the meaning of the actions of medieval mystics must be understood in terms of the religious idioms of their time and that their self-starvation represents not self-hatred, as suggested by literature on anorexia,[5] but a celebration of the power of suffering flesh (cited in Benson 1997: 138). In a very similar manner, my friends indeed celebrate their power over their flesh as an action of control that is reflective of the specific circumstance of independent Macedonia.

The grandmothers were far more critical of socialist Yugoslavia than their daughters, though it took time and trust before they would admit it. They felt that many things had been taken away from them and that socialism had been just another form of occupation. For them, beautification was a protest against the socialist ideology. Many times grandmothers said that a woman was a woman and should look like one. Finally, they suggested, women had some time for themselves, hardship had not cut deep lines into their faces, so now their daughters should have time to make themselves feel special and feel what it means to be a woman. Feeling womanly was a strong theme throughout the interviews with the grandmothers' generation. They emphasised how hard it had been to feel special about oneself and that modern times had finally allowed this to happen. They hoped that their daughters and granddaughters enjoyed this special, mutually inclusive gift God had given them: motherhood and femininity. Finally, a woman had time to take care of her body and was not sent to work in the fields while she was pregnant. It bothered these women greatly when they saw that their daughters were not taking advantage of these new opportunities and were pushing themselves too hard.

> Working, working, working, that's all they know. I did this all my life and so did my mother. Why does God not give us a rest? Today things are supposed to be better. Babies are born healthy, but still women work and work; "For what?" I ask myself, "For what?" It is time men become men and women are allowed to be women.

The grandmothers feel their granddaughters have, in one sense, the right attitude in that they want their bodies to be healthy and beautiful. They feel that their granddaughters are acting out the luxury of enjoying one's body,[6] except they think most of the young women are thin and wish they would eat more. I was told that the desire to be slim amongst young women showed that they had never known hunger, otherwise they would appreciate food and a well-fed body. It was understood that the grandmothers' role was to feed the granddaughters enough food to enable them to gain an understanding of a healthy body.

Today, some of my friends sell Western cosmetic products to their colleagues. If they manage to sell a specific quota of products, the cosmetic company awards them free gifts. What is important about these products is that they are manufactured and are presented in, for example, little glass flacons with gold coloured lids. They cannot be compared with the corn flour masks their mothers mixed for themselves, creating beauty out of nothing. Indeed, this is the main stimulus for buying foreign beauty products. What my friends praise, while their mothers look upon it suspiciously, is the sudden diversity of cosmetic products that can now be found in Macedonia itself, when once these were just seen on television or in shops in the West. Comparing this diversity with what they had known in their adolescence and, indeed, with what they continue to use today, the middle-aged women in my interviews stressed the feeling of uniformity with which they had grown up. One woman pointed out that they had joked on the beaches of Croatia that they could recognise a Yugoslavian woman from a Polish woman, because Polish women then had their hair dyed silver blond as opposed to the burgundy red hair colour used by Yugoslavian women.

I went shopping with my friends, and even until the late 1980s, the most advanced form of marketing used on the packaging of chocolate or hair shampoo was the image of a lady with a hairstyle from the 1960s. Women on the covers of magazines wore mini-skirts, but the colours were so faded one would assume that these were actually relics of the early sixties. Today, the world offers different images. MTV or other shows where advertising is shown are only watched by the younger generation. Vogue and

Mademoiselle magazines are laid out in the body sculpture studios, where the average customer is in her twenties. These new images are quite different, with different shades and different shapes compared to Yugoslavian advertising. They do not serve a utilitarian purpose; instead they are an expression of a lifestyle. Here the world of my friends separates from the world of their mothers and grandmothers. The lifestyle demands that you "be yourself," be independent, be firm and strong-willed. It draws my friends away from the uniformity, which they see as their mothers' burden. For them, the world of Western glamour is a world of choice, and, in contrast to their mothers, my friends feel a right to such choices. When I asked about the difference they saw between Yugoslavian/Macedonian and European products, I was told "Everything is glossy and shiny; our things are all dull." In the way my friends formulated this difference, they also formulated a boundary. The dull world was the world of their mothers, which they once thought they had escaped in much the same way as their mothers had escaped the hardships of rural life. The glossy Western world was on their doorstep. However, political changes in Yugoslavia seemed to throw them back into the uniformity of their mothers' lives. Many mothers also understood this intersection of the political and personal in the same way. Consequently, images of beauty gained a far greater importance in the personal lives of my friends.

Imaginative interpretations and images excite and inspire cultural discourse, but they are not limited to mirroring reality (Bordo 1997: 185). The mothers of my friends had a different way of coping with uniformity. They sewed and made clothes. Aided by their mothers, they made clothes for their grandchildren from Burda, a German sewing magazine with pattern sheets. However, some women I talked to considered corn-masks and pharmacist-mixed perfume, mini-skirts from Burda patterns, and burgundy red hair, as their specific way of dealing with the sometimes humiliating effect of this uniformity. They felt this uniformity had denied them their femininity. I was told, "The only thing then I wanted was to smell good and to feel clean. It was so hard to get even sanitary pads." Today, one can get sanitary pads but for five times the price. Nevertheless, they were the first items my friends bought from the new duty-free shops. They are an item mentioned to me by all three generations as an explanation of how women are valued in society. The grandmothers told me about the disgrace they felt when they were young and were using straw or grass bundles wrapped in cloth. I was told of the change in women's rights under socialism. For example, although women could do the same work as men, the system failed to supply a sufficient number of sanitary towels, which limited women's ability to work. Even today, tampons are locked up under glass in the stores and sold in packs of 10 at an exorbitant price. When my friends can rarely find any cheap sanitary pads in the market, but only the very expensive Western kind in the duty-free glamour stores, where the sanitary pads are in the window next to Marlboro cigarettes, Chanel perfume, Ariel washing powder, and Siemens microwaves, this situation is shown in all its absurdity. Here it can be interpreted that it is the social body that constrains the way the physical body is perceived. The physical experience of the body, always modified by the social categories through which it is known, sustains a particular view of society (Douglas 1970: 93).

The sculpted body presents itself as clean and effortless. Side effects are the immense effort that is required to create a body, which fits the image of Vogue magazines and, because of the starvation dieting, some of my friends suffered serious health problems. In Yugoslavia, when my young friends entered university, the number of female engineering students was still about 50 per cent of all engineering students. Today the number of women at the faculty has dropped dramatically. However, the so-called egalitarian system that socialist Yugoslavia created also created the double burden carried by my friends' mothers. Only their visibility, not society's attitude towards women had changed.

Women were used as an icon in Yugoslavia, as they are today in Macedonia. Living expenses were so high in Yugoslavia that both partners in a family had to work, but only the women did the housework. However, today with socialism withering, women try to hold onto their legal rights, and, to a certain extent, they have been successful. The demands placed on women are now subtler as I have shown in this chapter. Women's rights are not sacrificed for the benefit of some nationalist scheme. Women are not portrayed as the child bearers of the nation; they are not held responsible for keeping the nation "clean" by keeping outsiders out. Women in Macedonia are still in charge of their reproductive rights. Nevertheless, their bodies are used by social discourse. One could easily conclude that the bodies of my friends demonstrate both for themselves and for the surrounding society that Macedonians are in control, that they are on an equal footing with the powerful world around them—Western Europe and North America.

In *Gender Trouble*, Butler reads the body as a text. I, however, would like to understand the bodies of my informants, not as text, but as a concrete and limiting location and as actual social practice. To think that my friends starve themselves for two reasons—agency and resistance—ignores that those terms are concretely embedded in the social world of Skopje, Macedonia, today. So it cannot be concluded that my informants starve themselves because of the disintegration of Yugoslavia, or that we can read the disintegration of Yugoslavia on their bodies. Rather, they use cultural images of "slenderness," images that are as Western as they are Macedonian, to be more like their perceptions of other young people in Western Europe and, with this, delineate a difference between themselves and their parents. By being slender, they do not hope to be guided into Western Europe, nor do they entertain the illusion that they can in this way transform Macedonia into Western Europe. They merely want to control their bodies to make a difference: a difference between themselves and their parents. The disintegration of Yugoslavia, however, does cause them to see this difference diminishing before their eyes because they do not have the same avenue to escape the world of their parents as they had in Yugoslavia.

NOTES

1. See Noller and Callan 1991; Apter 1990; and Coleman and Hendry 1990.
2. These are "Western" fruits and have been available regularly in the shops for only a year.
3. The US gave a limited number of green cards to countries of their choice. In Macedonia, a lottery company bought these.
4. I say "inaccessible" because for the parents to start a new life outside Macedonia would be far more difficult than for their children. Education and ideology and just the fact that they have built up a home in Macedonia would make it unlikely that the parents would leave Macedonia for a long period of time. Only one parent of a friend was living in Moscow at the time of the research and that was because of holding a political position.
5. See Orbach (1978), Chernin (1985), and Wolf (1992).
6. I heard a few elderly women talk about how they enjoy watching their little granddaughters take a bath, and many women would make remarks on how beautiful and soft the skin of many of my young friends was.

Chapter Seven

Conclusion

My emphasis on this group of urban informants, namely, young female engineers in Skopje, is a result partially of my intention of taking women's points of view during a time of great change. Rather than examining things usually considered, that is, men and their politics or the rural perspective (the countryside being perceived often as most vulnerable to change), I chose to consider the lives of young, professional, and urban women. During the war in Bosnia and the conflict between Croatia and Serbia, the media focused on how the politics of change were imposed on and exercised in the countryside. People stressed that the countryside was more prone to propaganda and violence, and when women were discussed in the context of recent events in former Yugoslavia, women were represented as victims of rape and oppressive politics. This violence toward rural women is a very sad and real aspect of life in the former Yugoslavia. In the film, *Before the Rain*, the conflict arises in the countryside with women as the first victims. Not denying the importance of understanding social change and crisis within rural communities, I meant to concentrate on the urban environment to determine how the disintegration of Macedonia influenced this specific population and if and how a process of the construction of a sense of cultural identity was taking place. I found that the sense of identity is constructed in an ongoing process, which, by implication, speaks of the contested nature of Macedonian culture itself.

My research differs from other studies done in Macedonia and the Balkan area in that, while most of them are interested in continuity and thus survey traditional practices, I am more interested in how my informants perceive and renegotiate this meeting of the past and present. The manner in which the past informs the present interests me less, although I do not disregard the past. Macedonians are represented, and often see themselves, as people without history and, as such, without grounds to formulate an "identity." This feeling, I believe, could be named a search for identity. Of course this search does not mean that Macedonians have no identity or no history, though some parties might be interested in arguing this. What I mean is that it seems that Macedonians continually search for what they could be because history has imposed on them this need to define themselves.

Moore (1988: 1) suggests that women have always been present in ethnographic accounts, primarily because of the strong emphasis in anthropology on kinship and marriage. This emphasis has resulted in viewing kinship and marriage as central for the organisation of gender. In my research, I wanted to go beyond such a conception and follow Moore (1988: 2) when she points out that the real problem about incorporating women into anthropology lies not at the level of empirical research but at the theoretical and analytical level. I hope to have demonstrated with my research that one can gain insight into a changing society through the eyes of women, their perceptions and concerns in these extreme times. Certainly, I cannot claim, from my small group of informants, to have a clear understanding of the whole of Macedonia, but I can grasp through their eyes an insight into the working of this society, an insight that is not centred on kinship and marriage only, but rather one that deals on a day-to-day basis with the political and economic environment surrounding both.

In my work, I suggest additional points of entry for the study of a society in change and crisis and for urban research. I suggest that my informants, in presenting themselves to the "Other," internally the Albanian and externally the "West," respond to the way they would like to control their future. The question is whether the binary of "us" and "them" as an abstract perception is really relevant. It is apparent, however, that my friends see their world as an ongoing discussion of what was, what is, and what could be. It is at this intersection where I believe we must consider the complex role women play in this dialogue, which is crucial to the discourse on the identity of Macedonian society. One must not forget that in a world where individuals are forced to negotiate lifestyle choices among a diversity of options (Giddens 1991: 5), the choices between the "West" and "Balkan," "us" and "them" are only a few of many.

I have found that my informants provide an excellent lens on a changing present precisely because they will not submit to any fixed position within their society. That is, they are presenting a gendered picture that goes beyond marriage and kinship. Indeed, kinship and marriage are contested grounds because of past and present changes within their society. Discussions in the preceding chapters embraced questions of what it means to be Macedonian, female, urban, and "Western." These are, however, binaries, and the other half of these binaries feeds into the definitions my informants have of these opposite terms, so that the "Balkan," the "Albanians," and the countryside and past become counter-references through which the graduates understand their world. The spatial and symbolic content and context of my friends' obsession with beauty, Western consumer goods, and their working lives are integral to their construction of an adult identity within independent Macedonia.

Traditionally, anthropologists have treated the past as the legitimising tool for the present in their scholarship, including in the numerous discussions about "imagined communities" and "invented traditions." I, however, explored the opposite, namely, the future. In this respect, I see the future to be the legitimising tool for the specific life choices my informants make. My argument is that, if the past made my friends what they are today, then it is also the notion they have of the future that transforms them into what they will become. This statement underlies an idea about globalisation that Appadurai describes as being disseminated through global advertising within the plethora of creative and culturally well-chosen ideas of consumer agency. He states, "The globalisation of culture is not the same as its homogenisation, but globalisation involves the use of a variety of instruments of homogenisation (armaments, advertising techniques, language hegemonies, clothing styles and the like), which are absorbed into local political and cultural economies." Further, he states his belief that the central feature of global culture today is the politics of the mutual effort of sameness and difference (Appadurai 1990: 307). Thus, he is introducing globalisation as a value system, a value system that I believe has been taken up by the young urban *Skopjanici*. As I introduced their lives in the Republic of Macedonia, I introduced their consumer agency, as well as the changing Macedonian political and cultural economy. What became apparent is that, today, Macedonians, specifically my informants, are involved in a highly political discourse of sameness and difference, and they define their future around such a concept.

The extent of change in Macedonia has been greater than most of my informants, and probably I myself, have yet realised. When you live through the changes, it is sometimes difficult to see them on a day-to-day basis, and only upon reflection from some distance do certain aspects become apparent. The change has happened in Macedonia not only on a local basis, but a dramatic change has happened on a global scale with Macedonia being only a small part of it. This change has led to the breakdown of stable group membership. Today it is not as easy to talk about "identity" as it was just a few years ago. Today we speak of the "identities" of a person. Identities today derive from a multiplicity of sources: nationality, ethnicity, gender, age, and education, for instance. Those identities overlap and sometimes oppose each other, and they create their own fragmented identity mix. However, the experience of conflicting identities for my friends is not something that is specific to Macedonia. Far from it, this experience makes my friends part of the general global society, specifically through this multiplicity of identities. Despite this though, my friends have a specific location in this world, and it links them to the society they live in. I have demonstrated in my research how this specific dynamic is conceptualised by the graduates and how they make sense of the social, political, and economic changes around them. Their view on their surrounding world gives an insight into the workings of an "interface" between the individual and her social setting. Their present social setting gives them an idea of who they are, who they want to become, how they relate to others, and how they can understand the world they live in. It explains to them how they are the same and how they are different. In this respect, their identity becomes a specific moment in time, unfixed, fluent, and negotiable.

Yugoslavia today is understood to be a region of war, conflict, and interethnic hatred, and, as I and many others watching this conflict argue, this hatred is senseless. The first time I came to Macedonia, it did not differ very much from the other republics of Yugoslavia, and, even today, I cannot see specific reasons why Macedonia was not involved in an ethnic war of some sort. There are some political reasons for the lack of war, such as wise political decisions, foreign interest, and perhaps no interest from Serbia. Nevertheless, I think one can argue that Macedonia, theoretically, could have been put through the same sort of terror that destroyed Bosnia and Yugoslavia as a whole. People had shared fifty years under Tito and had experienced, even though it was manipulated and enforced, unity within this framework. They had shared a very similar day-to-day life with others in Yugoslavia. People in Macedonia still talk about "our Adriatic sea." Why were people outside of Macedonia suddenly so different that they were ready to kill each other? Was it because they could not live with the difference of "the other," whoever or whatever this was?

As is visible throughout these chapters on Macedonia, this difference of "the other" gained great importance within its own borders, but also within Macedonia's relation to the outside world. If I were to describe the war in Yugoslavia, I would have to say that it is a war about sameness and difference. This region became the latest temporal point for this conflict, which obviously started far earlier. (We will leave the historians to ponder on the exact date of origin.) In Macedonia, as in the other republics of the dissolved Yugoslavia, the common discourse on unity was turned to the other extreme, a common discourse on difference. This is the red line going through all the chapters: the (pre)occupation of people with difference. What is visible throughout the chapters, however, is that definitions of difference are in direct relation with perceptions of the others' sameness, either of the threatening sameness between the Slavic Macedonians and the Muslim Albanians or the anticipated sameness with "Western Europe." Sameness and difference are expressed through the things that people use. When I appeared in Macedonia with twenty white cotton socks that my mother had given me, my friends despaired: I should have known that only Albanians wore white socks. So with each daily visit to my flat, my friends brought me a new pair of socks while slowly the white socks disappeared out of my closets. This story might seem ridiculous, but the pervasiveness of the attitudes it exposes requires analysis. What my friends construct here is a specific difference that is as social as it is symbolic.

White socks are a complex issue; do Europeans wear white socks? It was inconceivable for them that I should do something that only uncultivated people would. However, my friends were not taken by surprise when my Canadian husband appeared with white socks as well. Nevertheless, they had no issue with his wearing white socks because, as I was told, everybody knows how uncultivated Americans, and therefore Canadians, are. Despite this, my husband was not allowed into some cafés because of his socks.

Throughout the chapters when writing about the West, I referred specifically to Northern Europe. Despite the enormous economic help they provided, American UNPROFOR troops and American businesses were seen as the old enemy. The uncultured, uncivilised America that Tito created did still exist. What this means is that the assertion of what it means to be Macedonian is historically specific. Sameness and difference is located in a particular point of time, though this temporal divide is variable. In Macedonia, this temporal divide starts with Ottoman rule and not with the Slavic invasion of Macedonia, which in itself is worth exploring. What this portrays, however, is Macedonians not as proud invaders and their own people, but as *raya*, cattle of the Sultan, as the passive recipients of sameness and difference. The next temporal divide, then, is between the "Balkans" and "Europe." Within this divide, Yugoslavia as a political unit was still firmly entrenched in the Balkans, something that the graduates were sure they could challenge in the future. However, the opportunity of challenging Macedonia's "Balkan" status also happened, albeit temporarily, in a specific moment of time, when Ante Markovich equalised the Yugoslavian currency with the Deutsche Mark and consequently brought Europe, in a very real economic sense, to the Balkans. Only a year later this dream was taken away.

Despite the dominant discourse of difference, which my friends are perpetuating and are presented with, they do endorse the cultural confluence of a global lifestyle and are producing newly shared identities. The notion of social change that has influenced Macedonia's political life has altered the life of my informants, but not in a way that could be defined as "post-socialism," which would suggest that socialism is over and democracy is coming. In their discourse of sameness and difference, the political and social conceptions of sameness and difference become fused. Socialism is a counter reference point to the life of my friends as much as democracy is. Within, Macedonia finds itself in a dynamic discourse on retaining socialist values and endorsing the unity that Tito had offered Yugoslavia, while, at the same time, striving for political, economic, and social democracy; endorsing the idea of a future as "Europeans"; and leaving behind the Balkan world of Yugoslavia. Internally, this causes discord between the Slavic Macedonian and Albanian population, but, even here, through a rural past and the example of Bosnia, sameness cannot be escaped and the inheritance of Tito is, to a certain degree, upheld. The flip side of this similarity is that Macedonia, in its political discourse, does not seek to return to a lost past; it is a past for which they gratefully acknowledge Tito, but which they have left behind while moving closer to "Western Europe." As such, a circle is formed that defines Macedonia's future.

Macedonia's future is formed through the shared idea of its future that arises out of the circle described above. I define a shared ideal as shared assumptions and representations of what it means to be "Western European," Albanian, Macedonian, or from the Balkans. Those shared assumptions and representations construct ideas of sameness and difference.

I infer from my research that changes are not only taking place on a political or social scale of global meaning, but on a very local and personal level. In particular, these changes presented my friends with different places on a global scale from which they formed what they wanted to be.

Individuals live within a large number of different institutions, or what Pierre Bourdieu (1984) calls "fields". Those are, for example, family, friends, work, and educational settings. The graduates participate in these fields exercising many different choices in varying degrees, but each of them moves in a specific space or context. How my friends are at home is different from how they are at work and different from how they are with their friends. However, the home, for example, is not only the relationship between daughter and mother but also the locality where television is watched and specific messages are received. This message, then, is responded to in a specific way of shopping or in the body-sculpting studios. The result of the body-sculpting studios and shopping, combined with who they are at their workplace and the conflicts they encounter there, comes together with their experience of the place where their grandmothers live, the village, and the different stories they are told from the past. All these experiences and messages form my informants' understanding of how they would like to get married and how their adult life should look. This formative process, however, still happens in a specific cultural framework in which the past and the future serve as reference points. In each and every event and in each and every locality in time and space, my friends position themselves differently to others and to the multiplicity of their own identities, according to the specific person they are at this point in time and place.

My friends define themselves and what their future should be through difference. One can argue that this is the most drastic change in Macedonian society. In Yugoslavia, the dominant social and political discourse was based on sameness, and today it is difference that prevails. My friends' subjectivity is forged through the marking of difference. This difference is related to what "Macedonians" are not, what they do not want to be, and what they wish to be.

I have presented here, obviously, only a very specialised ethnographic view of a small group of people. I have left out "the Other," that is, the Albanian and Roma populations. This omission is directly linked to my strong "kin" connection to the Slav Macedonians. Had I tried to step over this boundary, I would have broken the trust of the Macedonians, and it would have been very difficult for me to earn the trust of the Albanian and Roma population. Nevertheless, they were always present precisely because they were the counter reference my informants used to define who they are. However, I do not wish to confirm stereotypes of the Albanian and Roma population, and I also do not want to portray the Slav Macedonians as racist. The line between sameness and difference is very fine, and it is because of this circumstance that sameness and difference are so ferociously defended. This conflict between sameness and difference did not arise with the independence of Macedonia. It was smouldering all along in Yugoslavia, but the accentuation of it, the focus, has changed. Maybe it is just that the world in Macedonia became smaller, and the Albanian and Roma could not be overlooked, whereas, perhaps before, the gaze was directed towards Slovenia or Belgrade or Paris. As the world became smaller, fear and mistrust started, but this smallness also offered a chance to acknowledge "the Other" and to deal with the problem of "ethnic" hatred. Despite these pessimistic undertones, Macedonia has done extraordinarily well to stay out of the Yugoslav conflict, and the country has forged a solid path to independence.

I certainly do hope that all Macedonians in the Republic of Macedonia will have a future, and they will not be waiting in vain. Perhaps the future lies in recognising that sameness and difference are ultimately connected with each other. So *T'ga za Jug*, "Longing for the South," and "Waiting for Macedonia" combine the two aspects of the lives of my friends. First, comes the acceptance of who they are and where they are, the acceptance of their rural past and the acceptance of the Balkans as their place of origin, a place that has also reared Yugoslavia. Second, they are "Waiting for Macedonia" and wishing to be accepted by "Western Europe," to be the same as other young female engineering graduates in Paris, London, or Berlin and to have the same future as they have. The combination of these two aspects is their dilemma. These two notions, ranging from the *T'ga za Jug* to "Waiting for Macedonia," are what moves the lives of my friends, defines who they are, defines their subjectivity, and eventually, what the Republic of Macedonia is today.

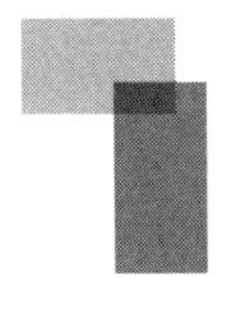

Bibliography and Recommended Reading

Abrahams, R. 1996. *After Socialism: Land Reform and Rural Social Change in Eastern Europe.* Oxford: Berghahn Books.

Abu-Lughod, L. 1986. *Veiled Sentiments: Honor and Poetry in a Bedouin Society.* Berkeley: University of California Press.

Abu-Lughod, L. 1993. *Writing Women's Worlds: Bedouin Stories.* Berkeley: University of California Press.

Adam, B. 1995. *Timewatch: The Social Analysis of Time.* Cambridge: Polity Press.

Allen, H.B. 1943. *Come Over Into Macedonia: The Story of a Ten-Year Adventure in Uplifting a War-Torn People.* New Brunswick: Rutgers University Press.

Althusser, L. 1971. *Lenin and Philosophy and Other Essays.* Trans. Ben Brewster. London: NLB.

Amit-Talai, V. and H. Wulff. 1995. *Youth Cultures: A Cross-Cultural Perspective.* London: Routledge.

Anderson, B. [1983] 1991. *Imagined Communities: Reflections on the Origins and Spread of Nationalism.* London: Verso.

Anderson, K. and F. Gale, eds. 1992. *Inventing Places: Studies in Cultural Geography.* Melbourne: Longman Cheshire.

Andreev, V. and L. Jakimovski. 1993. *The Republic of Macedonia.* Skopje: Goce Delcev.

Andrejevich, M. 1991. Resurgent Nationalism in Macedonia: A Challenge to Pluralism. *Report on Eastern Europe* 2: 26–29.

Andrews, E. 1981. *Closing the Iron Cage: The Scientific Management of Work and Leisure.* Montreal: Black Rose Books.

Andrusz, G. 1992. Housing Reform and Social Conflict. In *Russia in Flux: The Political and Social Consequences of Reform,* ed. D. Lane, 210–38. London: Edward Elgar.

Anthony, P.D. 1977. *The Ideology of Work.* London: Tavistock.

Appadurai, A. 1986. Introduction: Commodities and the Politics of Value. In *The Social Life of Things: Commodities in Cultural Perspective,* ed. A. Appadurai, 3–63. Cambridge: Cambridge University Press.

Appadurai, A., ed. 1986. *The Social Life of Things: Commodities in Cultural Perspective.* Cambridge: Cambridge University Press.

Appadurai, A. 1990. Disjuncture and Difference in the Global Cultural Economy. In *Theory, Culture and Society*, 7: 295–310.

Appadurai, A. 1996a. *Modernity at Large: Cultural Dimensions of Globalization.* Vol. 1 of *Public Works.* Minneapolis: University of Minnesota Press.

Appadurai, A. 1996a. Sovereignty Without Territoriality: Notes for a Post-national Geography. In *The Geography of Identity*, ed. P. Yaeger, 40–58. Ann Arbor: University of Michigan Press.

Applebaum, H., ed. 1984. *Work in Non-Market and Transitional Societies.* SUNY Series in the Anthropology of Work. Albany: State University of New York Press.

Applebaum, H. 1992. *The Concept of Work: Ancient, Medieval, and Modern.* SUNY Series in the Anthropology of Work. Albany: State University of New York Press.

Applebaum, H. 1995. The Concepts of Work in Western Thought, trans. J.C. Nash. In *Meanings of Work: Considerations for the Twenty-First Century*, ed. F.C. Gamst, 46–78. SUNY Series in the Anthropology of Work. Albany: State University of New York Press.

Apter, E.S. and W. Pietz. 1993. *Fetishism as Cultural Discourse.* Ithaca: Cornell University Press.

Apter, T.E. 1990. *Altered Loves: Mothers and Daughters During Adolescence.* Hemel Hempstead: Harvester Wheatsheaf.

Ardener, S. 1975. *Perceiving Women.* London: Malaby Press.

Ardener, S. [1978] 1993. *Defining Females: The Nature of Women in Society.* Oxford: Berg.

Argyrou, V. 1996. *Tradition and Modernity in the Mediterranean: The Wedding as Symbolic Struggle.* Cambridge Studies in Social and Cultural Anthropology, No. 101. New York: Cambridge University Press.

Armstrong, T. 1998. *Modernism, Technology, and the Body: A Cultural Study.* Cambridge: Cambridge University Press.

Atkinson, D. 1977. Society and the Sexes in the Russian Past. In *Women in Russia*, ed. D. Atkinson, Alexander Dallin and G.W. Lapidus, 3–38. Stanford: Stanford University Press.

Atkinson, J.M. 1989. *The Art and Politics of Wana Shamanship.* Berkeley: University of California Press.

Austin, J. and M. Willard. 1998. *Generations of Youth: Youth Cultures and History in Twentieth-Century America.* New York: New York University Press.

Axt, H.J. 1993. Macedonia: Struggle Over Names or Conflict Before the Eruption. *Europa Archiv* 48: 65–75.

Baker, A.R.H. 1992. On Ideology and Landscape. In *Ideology and Landscape in Historical Perspective: Essays on the Meanings of Some Places in the Past*, ed. A.R.H. Baker and G. Biger. Cambridge Studies in Historical Geography, No. 18. Cambridge: Cambridge University Press.

Baker, A.R.H. and G. Biger, eds. 1992. *Ideology and Landscape in Historical Perspective: Essays on the Meanings of Some Places in the Past.* Cambridge Studies in Historical Geography, No. 18. Cambridge: Cambridge University Press.

Balkan Forum: An International Journal of Politics, Economics and Culture. 1995. 3.

Banac, I. 1984. *The National Question in Yugoslavia: Origins, History, Politics.* Ithaca: Cornell University Press.

Baric, L. 1967a. Traditional Groups and New Economic Opportunities in Rural Yugoslavia. In *Themes in Economic Anthropology*, ed. R. Firth, 253–81. London: Tavisstock.

Baric, L. 1967b. Levels of Change in Yugoslav Kinship. In *Social Organization: Essays Presented to Raymond Firth*, ed. M. Freedman and R. Firth, 1–24. London: Cass.

Barth, F., ed. 1969. *Ethnic Groups and Boundaries.* Boston: Little Brown.

Barth, F. 1989. The Analysis of Culture in Complex Societies. *Ethnos* 3: 120–42.

Battaglia, D. 1995. *Rhetorics of Self-Making.* Berkeley: University of California Press.

Bebel, A. 1891. *Die Frau Und Der Sozialismus.* Stuttgart: Dietz Verlag.

Before the Rain. 1994. Directed by Milcho Mancheuski. Macedonia/France/UK.

Bell, P. 1984. *Peasants in the Socialist Transition.* Berkeley: University of California Press.

Belmonte, T. 1989. *The Broken Fountain.* New York: Columbia University Press.

Bender, B., ed. 1993. *Landscape: Politics and Perspectives.* Explorations in Anthropology Series. Oxford: Berg.

Benson, S. 1997. The Body, Health and Eating Disorders. In *Identity and Difference*, ed. K. Woodward, 122–44. London: Sage in Association with the Open University.

Benton, L.A. 1990. *Invisible Factories: The Informal Economy and Industrial Development in Spain.* SUNY Series in the Anthropology of Work. Albany: State University of New York Press.

Berend, I. 1986. The Crisis Zone of Europe: An Interpretation of East-Central European History in the First Half of the Twentieth Century. Cambridge; University Press.

Bloch, M. 1981. Descent and Sources of Contradiction in Representations of Women and Kinship. In *Gender and Kinship: Essays Toward a Unified Analysis*, ed. J.F. Collier and S.J. Yanagisako, 324–37. Stanford: Stanford University Press.

Bloch, M. 1983. *Marxism and Anthropology: The History of a Relationship.* Oxford: Clarendon.

Bloch, M. 1986. *From Blessing to Violence.* London: Cambridge University Press.

Bloch, M. 1989. *Ritual, History and Power: Selected Papers in Anthropology.* London School of Economics Monographs on Social Anthropology, No. 58. London: Athlone Press.

Bloch, M. and J. Parry, eds. 1982. *Death and the Regeneration of Life.* Cambridge: Cambridge University Press.

Bloch, M. and J. Parry, eds. 1989. *Money and the Morality of Exchange.* Cambridge: Cambridge University Press.

Bloch, M.E.F. 1998. *How We Think They Think: Anthropological Approaches to Cognition, Memory, and Literacy.* Boulder: Westview Press.

Boddy, J. 1989. *Wombs and Alien Spirits: Women, Men, and the Z'ar Cult in Northern Sudan.* New Directions in Anthropological Writing. Madison: University of Wisconsin Press.

Boehm, C. 1984. *Blood Revenge: The Anthropology of Feuding in Montenegro and Other Tribal Societies.* Lawrence: University Press of Kansas.

Boissevain, J. 1974. *Friends of Friends: Networks, Manipulations and Coalition.* Oxford: Blackwell.

Bordo, S. 1993. *Unbearable Weight: Feminism, Western Culture, and the Body.* Berkeley: University of California Press.

Bordo, S. 1997. *Twilight Zones: The Hidden Life of Cultural Images From Plato to O.J.* Berkeley: University of California Press.

Borneman, J. 1992. *Belonging in the Two Berlins: Kin, State, Nation.* Cambridge Studies in Social and Cultural Anthropology, No. 86. Cambridge: Cambridge University Press.

Bourdieu, P., ed. 1963. The Attitude of the Algerian Peasant Toward Time. *Mediterranean Countrymen: Essays in the Social Anthropology of the Mediterranean.* Paris: Mouton.

Bourdieu, P. 1984. *Distinction: A Social Critique of the Judgement of Taste.* London: Routledge and Kegan Paul.

Bourdieu, P. 1995. *Outline of a Theory of Practice.* Trans. Richard Nice. Cambridge Studies in Social and Cultural Anthropology, No. 16. Cambridge: Cambridge University Press.

Brailsford, H.N. [1906] 1971. *Macedonia: Its Races and Their Future.* London: Methuen.

Brandes, S. 1981. Like Wounded Stags: Male Sexual Ideology in an Andalusian Town. In *Sexual Meanings: The Cultural Construction of Gender and Sexuality,* ed. S.B. Ortner and H. Whitehead, 216–38. Cambridge: Cambridge University Press.

Brewster, B.R. and M. Mauss. 1979. *Sociology and Psychology: Essays.* London: Routledge and Kegan Paul.

Bridger, S., R. Kay and K. Pinnick. 1996. *No More Heroines?: Russia, Women and the Market.* London: Routledge.

Brima Gallup Report 1994.

Bringa, T. 1995. *Being Muslim the Bosnian Way: Identity and Community in a Central Bosnian Village.* Princeton Studies in Muslim Politics. Princeton: Princeton University Press.

Bringa, T.R. 1991. Gender, Religion and the Person: The "Negotiation" of Muslim Identity in Rural Bosnia. Ph.D. diss., University of London.

British Macedonian Social Surveys. 1994. *Political and Economical Index.* Brima in Joint Venture With Gallup Report 0194.

Brown, K. 1995. Of Meanings and Memories: The National Imagination in Macedonia. Ph.D. diss., The University of Chicago.

Brown, K.S. 1994. Seeing Stars: Character and Identity in the Landscape of Modern Macedonia. *Antiquity* 68: 786–96.

Browning, G. and A. Wason. 1992. Perestroika and Female Politicization. In *Russia in Flux: The Political and Social Consequences of Reform,* ed. D. Lane, 166–84. London: Edward Elgar.

Buckley, M. 1989. *Women and Ideology in the Soviet Union.* London: Harvester.

Bulag, U.E. 1998. *Nationalism and Hybridity in Mongolia.* Oxford Studies in Social and Cultural Anthropology. Oxford: Clarendon Press.

Burbach, R., O. Núñez Soto and B. Kagarlitsky. 1997. *Globalization and Its Discontents: The Rise of Postmodern Socialisms.* Chicago: Pluto Press.

Butler, J. 1990. *Gender Trouble: Feminism and the Subversion of Identity.* London: Routledge.

Butler, J.P. 1993. *Bodies That Matter: On the Discursive Limits of "Sex."* London: Routledge.

Bynum, C.W. 1987. *Holy Feast and Holy Fast: The Religious Significance of Food to Medieval Women.* The New Historicism. Berkeley: University of California Press.

Byrnes, R., ed. 1976. *Communal Families in the Balkans: The Zadruga.* Notre Dame, IN: University of Notre Dame Press.

Cabezali, E., M. Cuevas and M.T. Chicote. 1990. Myth as Suppression: Motherhood and the Historical Consciousness of the Women of Madrid, 1936–39. In *The Myths We Live By,* ed. R. Samuel and P. Thompson, 161–73. London: Routledge.

Campbell, J. 1989. *Joy in Work, German Work: The National Debate, 1800–1945.* Princeton, NJ: Princeton University Press.

Cannell, F. 1992. Catholicism, Spirit Mediums and the Ideal of Beauty in a Bicolano Community, Philippines. Ph.D. diss., University of London.

Cannell, F. 1995. The Power of Appearances: Beauty, Mimicry and Transformation in Bicol. In *Discrepant Histories: Translocal Essays on Filipino Cultures,* ed. V.L. Rafael, 223–58. Asian American History and Culture Series. Philadelphia: Temple University Press.

Caplan, P., ed. 1987. *The Cultural Construction of Sexuality.* London: Tavistock.

Caplan, P. and J. Bujra, eds. 1978. *Women United, Women Divided.* London: Tavistock.

Carrithers, M., S. Collins and S. Lukes, eds. 1985. *The Category of the Person: Anthropology, Philosophy, History.* Cambridge: Cambridge University Press.

Charles, N. 1993. *Gender Divisions and Social Change.* Hemel Hempstead, Hertfordshire: Barnes and Noble Books.

Chernin, K. 1982. *The Obsession: Reflections on the Tyranny of Slenderness.* New York: Harper Colophon Books.

Chernin, K. 1985. *The Hungry Self: Women, Eating and Identity.* New York: Harper and Row.

Chloros, A.G. 1979. *Yugoslav Civil Law: History, Family, Property.* Oxford: Clarendon Press.

Clarke, J. and C. Critcher. 1985. *The Devil Makes Work.* Urbana: University of Illinois Press.

Clemitson, I. and G. Rogers. 1981. *A Life to Live: Beyond Full Employment.* London: Junction Books.

Cleverley, J.F. 1985. *The Schooling of China: Tradition and Modernity in Chinese Education.* Sydney: Allen and Unwin.

Clifford, J. 1983. On Ethnographic Authority. *Representations* 1: 118–46.

Clifford, J. and G. Marcus, eds. 1986. *Writing Culture: The Poetics and Politics of Ethnography.* Berkeley: University of California Press.

Cohen, L. 1993. *Broken Bonds: The Disintegration of Yugoslavia.* Boulder: Westview Press.

Cohen, P.A. 1987. *Between Tradition and Modernity: Wang T'ao and Reform in Late Ch'ing China.* Harvard East Asian Monographs, No. 133. Cambridge, MA: Council on East Asian Studies Harvard University.

Cole, S. 1991. *Women of the Praia: Work and Lives in a Portuguese Coastal Community.* Princeton: Princeton University Press.

Coleman, J.C. and L.B. Hendry. 1990. *The Nature of Adolescence.* Adolescence and Society series. London: Routledge.

Collier, J.F., S.J. Yanagisako and M. Bloch. 1987. *Gender and Kinship: Essays Toward a Unified Analysis.* Stanford: Stanford University Press.

Comaroff, J. 1985. *Body of Power, Spirit of Resistance: The Culture and History of a South African People.* Chicago: University of Chicago Press.

Comaroff, J.L. 1980. *The Meaning of Marriage Payments.* Studies in Anthropology, No. 7. New York: Academic Press.

Comaroff, J.L. and J. Comaroff. 1992. *Ethnography and the Historical Imagination.* Studies in the Ethnographic Imagination. Boulder: Westview Press.

Conboy, K., N. Medina and S. Stanbury. 1997. *Writing on the Body: Female Embodiment and Feminist Theory.* Gender and Culture Reader. New York: Columbia University Press.

Cornwall, A. and N. Lindisfarne, eds. 1993. *Dislocating Masculinity.* Comparative Ethnographies. London: Routledge.

Corrin, C., ed. 1992. *Superwomen and the Double Burden: Women's Experience of Change in Central and Eastern Europe and the Former Soviet Union.* London: Scarlet Press.

Corrin, C. and J. Campling. 1994. *Magyar Women: Hungarian Women's Lives, 1960s-1990s.* Basingstoke: Macmillan; New York: St. Martin's Press.

Cowan, J.K. 1990. *Dance and the Body Politic in Northern Greece.* Princeton: Princeton University Press.

Cowan, J.K. 1991. Going Out for Coffee? Contesting the Grounds of Gendered Pleasures in Everyday Sociability. In *Contested Identities: Gender and Kinship in Modern Greece,* ed. P. Loizos and E. Papataxiarchis, 180–202. Princeton: Princeton University Press.

Crawford, R. 1984. A Cultural Account of "Health": Control, Release, and the Social Body. In *Issues in the Political Economy of Health Care,* ed. J.B. McKinlay, 60–103. London: Tavistock.

Crawford, S. 1985. Person and Place in Kalavasos: Perspectives on Social Change in a Greek-Cypriot Village. Ph.D. diss., University of Cambridge.

Csordas, T. 1988. Embodiment as a Paradigm for Anthropology. *Ethos* 18: 5–47.

Czyzewski, D. and M.A. Suhr, eds. 1988. *Conversations With Anorexics: Hilde Bruch.* New York: Basic Books

Dabac, T.S. and C. Grozdanov. 1961. *Ohrid.* Belgrade: Jugoslavija.

Dalton, G., ed. 1971. *Primitive, Archaic, and Modern Economics: Essays of Karl Polanyi.* Boston: Beacon Press.

Danforth, L.M. 1993. Claims to Macedonian Identity: The Macedonian Question and the Breakup of Yugoslavia. *Anthropology Today* 9: 3–10.

Danforth, L.M. 1995. *The Macedonian Conflict: Ethnic Nationalism in a Transnational World.* Princeton: Princeton University Press.

Danopoulos, C.P. and K.G. Messas. 1997. *Crises in the Balkans: Views From the Participants.* Boulder: Westview Press.

Davis, J., T.A. Hirschl and M. Stack. 1997. *Cutting Edge: Technology, Information Capitalism and Social Revolution.* London: Verso.

Davis, K. 1997. *Embodied Practices: Feminist Perspectives on the Body.* London: Sage.

Davis, K., M. Leijenaar and J. Oldersma. 1991. *The Gender of Power.* London: Sage.

Del Valle, T. 1993. *Gendered Anthropology.* London: Routledge.

Delaney, C. 1991. *The Seed and the Soil: Gender and Cosmology in Turkish Village Society.* Berkeley: University of California Press.

Denich, B. 1974. Sex and Power in the Balkans. In *Women, Culture and Society,* ed. M.Z. Rosaldo and L. Lamphere, 243–62. Palo Alto: Stanford University Press.

Denich, B. 1977. Women, Work and Power in Modern Yugoslavia. In *Sexual Stratification: A Cross-Cultural View*, ed. A. Schlegel, 215–44. New York: Columbia University Press.

Diemberger, H., G. Hazod and C. Schicklgruber. 1989. Mutterwort Und Vaterfolge: Frauen Und Mütter Machen Gesellschaft: Das Beispiel Der Khumbo. In *Von Fremden Frauen*, 325–88. Frankfurt/Main: Suhrkamp.

Dikoetter, F. 1995. *Sex, Culture and Modernity in China*. London: C. Hurst and Co. Ltd.

Dimitrova, B. 1993. Language and Politics in Bulgaria. In *Towards a New Community: Culture and Politics in Post-Totalitarian Europe*, ed. P.J.S. Duncan and M. Rady, 133–45. Hamburg and Münster: LIT Verlag.

Dixon, S. 1993. What Price an Orthodox Revival? The Dilemmas of the Russian Church. In *Towards a New Community: Culture and Politics in Post-Totalitarian Europe*, ed. P.J.S. Duncan and M. Rady. Hamburg and Münster: LIT Verlag.

Djilas, A. 1993. A Paper House: The Ending of Yugoslavia. *New Republic* 208: 38–42.

Douglas, M. 1966. *Purity and Danger: An Analysis of Concepts of Pollution and Taboo*. New York: Praeger.

Douglas, M. 1970. *Natural Symbols: Explorations in Cosmology*. London: Barrie and Rockliff Cresset Press.

Douglas, M. 1975. *Implicit Meanings: Essays in Anthropology*. London: Routledge and Paul.

Dragadze, T. 1993. The Domestication of Religion Under Soviet Communism. In *Socialism: Ideals, Ideologies and Local Practice*, ed. C.M. Hann, 148–56. London: Routledge.

Du Boulay, J. 1986. Women: Images of Their Nature and Destiny in Rural Greece. In *Gender and Power in Rural Greece*, ed. J. Dubisch, 139–68. Princeton: Princeton University Press.

Du Boulay, J. [1974] 1994. *Portrait of a Greek Mountain Village*. Limni (Evia), Greece: Denise Harvey.

Dubisch, J. 1986. *Gender and Power in Rural Greece*. Princeton: Princeton University Press.

Dubisch, J. 1991. Gender, Kinship, and Religion: "Reconstructing" the Anthropology of Greece. In *Contested Identities: Gender and Kinship in Modern Greece*, ed. P. Loizos and E. Papataxiarchis, 33–46. Princeton: Princeton University Press.

Dubisch, J. 1995. *In a Different Place: Pilgrimage, Gender and Politics at a Greek Island Shrine*. Princeton Modern Greek Studies. Princeton: Princeton University Press.

Duncan, P.J.S. and M. Rady, ed. 1993. *Towards a New Community: Culture and Politics in Post-Totalitarian Europe*. Hamburg and Münster: LIT Verlag.

Dwyer, K. 1977. The Dialogic of Anthropology. *Dialectical Anthropology* 2: 143–51.

Dwyer, K. 1982. *Moroccan Dialogues.* Baltimore: John Hopkins University Press.

Einhorn, B. 1993. *Cinderella Goes to the Market: Citizenship, Gender and Women's Movements in East Central Europe.* London: Verso.

Elshtain, J.B. 1981. *Public Man, Private Woman: Women in Social and Political Thought.* Princeton: Princeton University Press.

Engels, F. [1884] 1972. *The Origin of the Family, Private Property, and the State.* New York: Pathfinder Press.

Erlich, V.S. 1966. *Family in Transition: A Study of 300 Yugoslav Villages.* Princeton: Princeton University Press.

Errington, F. and D. Gewertz. 1987. *Cultural Alternatives and a Feminist Anthropology: An Analysis of Culturally Constructed Gender Interests in Papua New Guinea.* Cambridge: Cambridge University Press.

Errington, R.M. 1990. *A History of Macedonia.* Hellenistic Culture and Society, No. 5. Berkeley: University of California Press.

Fabian, J. 1983. *Time and the Other: How Anthropology Makes Its Object.* New York: Columbia University Press.

Falk, P. and C. Campbell. 1997. *The Shopping Experience.* Theory, Culture and Society. London: Sage.

Favret-Saada, J. 1980. *Deadly Words: Witchcraft in the Bocage.* Cambridge: Cambridge University Press.

Featherstone, M. 1990. Global Culture: An Introduction. In *Global Culture: Nationalism, Globalization, and Modernity,* ed. M. Featherstone, 1–14. A Theory, Culture and Society Special Issue. London: Sage.

Featherstone, M., ed. 1990. *Global Culture: Nationalism, Globalization, and Modernity.* London: Sage.

Featherstone, M. 1995. *Undoing Culture: Globalization, Postmoderism and Identity.* Theory, Culture and Society. London: Sage.

Feher, M., R. Naddaff and N. Tazi. 1989. *Fragments for a History of the Human Body.* 3 vols. Zone, No. 3–5. New York: Zone Books.

Field, N. 1995. The Child as Laborer and Consumer: The Disappearance of Childhood in Contemporary Japan. In *Children and the Politics of Culture,* ed. S. Stephens, 51–76. Princeton: Princeton University Press.

Filipovic, M.S. 1976. Zadruga (Kac'na [sic] zadruga). In *Communal Families in the Balkans,* ed. R. Byrnes, 268–79. Notre Dame: University of Notre Dame Press.

Firlit, E. and J. Chlopecki. 1992. When Theft Is Not Theft. In *The Unplanned Society: Poland During and After Communism.* New York: Columbia University Press.

Firth, R., ed. 1967. *Themes in Economic Anthropology.* London: Tavistock.

Firth, R. 1972. *The Organization of Work Among the New Zealand Maori.* Wellington, New Zealand: Shearer.

Ford, G. 1983. "Za Dusha": An Interpretation of Funeral Practices in Macedonia. *Symbolic Interaction* 6: 19–34.

Ford, G.H. 1982. Networks, Ritual and Vrski: A Study of Urban Adjustment in Macedonia. Ph.D. diss., Arizona State University.

Foucault, M. 1978. *The History of Sexuality*. Vol. 1. London: Allen Lane.

Foucault, M. and A. Sheridan. [1975] 1995. *Discipline and Punish: The Birth of the Prison*. New York: Vintage Books.

Franklin, S. 1997. *Embodied Progress: A Cultural Account of Assisted Conception*. London: Routledge.

Freedman, M. and R. Firth. 1967. *Social Organization: Essays Presented to Raymond Firth*. London: Cass.

Friedl, E. 1962. *Vasilika: A Village in Modern Greece*. New York: Holt, Rinehart and Winston.

Friedl, E. [1967] 1986. The Position of Women: Appearance and Reality. In *Gender and Power in Rural Greece*, ed. J. Dubisch, 42–52. Princeton: Princeton University Press.

Friedman, J. 1990. Being in the World: Globalization and Localization. In *Global Culture: Nationalism, Globalization, and Modernity*, ed. M. Featherstone, 311–28. A Theory, Culture and Society Special Issue. London: Sage.

Gamst, F.C. 1993. "On Time" and the Railroader: Temporal Dimensions of Work. In *Ethnologie Der Arbeitswelt*, ed. S. Helmers, 105–31. Bonn: Holos.

Gamst, F.C. 1995. Considerations of Work, trans. J.C. Nash. In *Meanings of Work: Considerations for the Twenty-First Century*, ed. F.C. Gamst. SUNY Series in The Anthropology of Work. Albany: State University of New York Press.

Gamst, F.C., ed. 1995. *Meanings of Work: Considerations for the Twenty-First Century*. SUNY Series in The Anthropology of Work. Albany: State University of New York Press.

Gamst, F.C. and S. Helmers. 1991. Die Kulturelle Perspektive Und Die Arbeit: Ein Forschungsgeschichtliches Panorama Der Nordamerikanischen Industrieethnologie. *Zeitschrift Für Ethnologie* 116: 25–41.

Gans, H.J. 1962. *The Urban Villagers: Group and Class in the Life of Italian Americans*. New York: Free Press.

Gardner, K. 1995. *Global Migrants, Local Lives: Travel and Transformation in Rural Bangladesh*. Oxford: Clarendon Press.

Gell, A. 1986. Newcomers to the World of Goods: Consumption Among the Muria Gonds. In *The Social Life of Things: Commodities in Cultural Perspective*, ed. A. Appadurai, 110–38. Cambridge: Cambridge University Press.

Gell, A. 1992. *The Anthropology of Time: Cultural Constructions of Temporal Maps and Images*. Oxford: Berg.

Gellner, E. 1983. *Nations and Nationalism*. Oxford: Blackwell.

Gennep, A.V. 1960. *The Rites of Passage.* Chicago: University of Chicago Press.

Gibson, T. 1986. *Sacrifice and Sharing in the Philippine Highlands: Religion and Society Among the Buid of Mindoro.* London: Athlone Press.

Giddens, A. 1979. *Central Problems in Social Theory: Action, Structure and Contradiction in Social Analysis.* London and Basingstoke: Macmillan.

Giddens, A. 1990. *The Consequences of Modernity.* Stanford: Stanford University Press.

Giddens, A. 1991. *Modernity and Self-Identity: Self and Society in the Late Modern Age.* Stanford: Stanford University Press.

Giddens, A. 1995. *Politics, Sociology and Social Theory: Encounters With Classical and Contemporary Social Thought.* Stanford: Stanford University Press.

Giddens, A. [1992] 1995. *The Transformation of Intimacy: Sexuality, Love and Eroticism in Modern Societies.* Cambridge: Polity Press.

Glenny, M. 1993. *The Rebirth of History: Eastern Europe in the Age of Democracy.* Harmondsworth: Penguin.

Gligorov, K. 1995. Macedonian Model of Peace and Security in the Balkans. *Balkan Forum* 3: 344.

Goffman, E. 1972. *Interaction Ritual.* Harmondsworth: Penguin.

Goldman, W.Z. 1993. *Women, the State and Revolution: Soviet Family Policy and Social Life, 1917–1936.* Cambridge: Cambridge University Press.

Goldschmidt, W. 1995. Task Performance and Fulfilment: Work and Career in Tribal and Peasant Societies, trans. J.C. Nash. In *Meanings of Work: Considerations for the Twenty-First Century,* ed. F.C. Gamst. SUNY Series in The Anthropology of Work. Albany: State University of New York Press.

Goody, J. 1976. *Productions and Reproduction: A Comparative Study of the Domestic Domain.* Cambridge: Cambridge University Press.

Goody, J. 1983. *The Development of the Family and Marriage in Europe.* Past and Present Publications. Cambridge: Cambridge University Press.

Goody, J. and S.J. Tambiah. 1973. *Bridewealth and Dowry.* Cambridge Papers in Social Anthropology, No. 7. Cambridge: Cambridge University Press.

Gorz, A. 1982. *Farewell to the Working Class.* London: Pluto Press.

Gorz, A. 1985. *Paths to Paradise: On the Liberation From Work.* Boston: South End Press.

Gorz, A. 1986. The Socialism of Tomorrow. *Telos* 67: 199–205.

Götz, I. and A. Moosmüller. 1992. Zur Ethnologischen Erforschung Von Unternehmenskulturen. Industriebetriebe Als Forschungsfeld Der Völker Und Volkskunde. *Schweizerisches Archiv Für Volkskunde* 88: 1–30.

Hall, R.H. 1986. *Dimensions of Work.* Beverly Hills: Sage.

Halpern, J.M. 1963. Yugoslav Peasant Society in Transition: Stability in Change. *Anthropological Quarterly* 36: 156–82.

Halpern, J.M. 1972. *A Serbian Village in Historical Perspective.* New York: Holt, Rinehart and Winston.

Halpern, J.M. and D. Anderson. 1970. The Zadruga: A Century of Change. *Anthropologica* 12: 83–97.

Halpern, J.M. and B. Halpern. 1979. Changing Perceptions of Roles as Husbands and Wives in Five Yugoslav Villages. In *Europe as a Cultural Area*, ed. J. Cuisenier, 159–72. Chicago: Aldine.

Halpern, J.M. and D. Kideckel. 1983. Anthropology of Eastern Europe. *Annual Review of Anthropology* 12: 377–402.

Hammel, E.A. 1968. *Alternate Social Structures and Ritual Relations in the Balkans.* New Jersey: Prentice Hall.

Hammel, E.A. 1969. Economic Change, Social Mobility, and Kinship in Serbia. *Southwestern Journal of Anthropology* 25: 188–97.

Hammel, E.A. 1972. The Zadruga as Process. In *Household and Family in Past Time*, ed. P. Laslett and R. Wall, 335–73. Cambridge: Cambridge University Press.

Hammel, E.A. 1995. Economics 1, Culture 0: Fertility Change and Differences in the Northwest Balkans, 1700–1900. In *Situating Fertility: Anthropology and Demographic Inquiry*, ed. S. Greenhalgh, 225–58. Cambridge: Cambridge University Press.

Hammel, E.A. and C. Yarbrough. 1973. Social Mobility and the Durability of Family Ties. *Journal of Anthropological Research* 29: 145–63.

Hammond, N.G.L. 1979. *A History of Macedonia: Volume II. 550–336 B.C.* Oxford: Oxford University Press.

Hammond, N.G.L. 1989. *The Macedonian State: Origins, Institutions, and History.* Oxford: Clarendon Press.

Handler, R. 1988. *Nationalism and the Politics of Culture in Quebec.* Madison: University of Wisconsin Press.

Hann, C.M., ed. 1993. *Socialism: Ideals, Ideologies and Local Practice.* London: Routledge.

Hareven, T. 1982. *Family Time and Industrial Time.* Cambridge: Cambridge University Press.

Heilbrun, C.G. 1988. *Writing a Woman's Life.* New York: Ballantine.

Hendry, J. 1993. *Wrapping Culture: Politeness, Presentation and Power in Japan and Other Societies.* Oxford Studies in the Anthropology of Cultural Forms. Oxford: Clarendon Press.

Hendry, J. 1995. *Understanding Japanese Society.* The Nissan Institute/ Routledge Japanese Studies Series. London: Routledge.

Hendry, J. 1998. *Interpreting Japanese Society: Anthropological Approaches.* London: Routledge.

Herzfeld, M. 1982. *Ours Once More: Folklore Ideology and the Making of Modern Greece.* Austin: University of Texas Press.

Herzfeld, M. 1985. *The Poetics of Manhood: Contest and Identity in a Cretan Mountain Village.* Princeton: Princeton University Press.

Herzfeld, M. 1987. *Anthropology Through the Looking Glass.* Cambridge: Cambridge University Press.

Hesse-Biber, S.J. 1996. *Am I Thin Enough Yet?: The Cult of Thinness and the Commercialization of Identity.* New York: Oxford University Press.

Hewitt, K. 1997. *Mutilating the Body: Identity in Blood and Ink.* Bowling Green, OH: Bowling Green State University Popular Press.

Heywood, L. 1998. *Bodymakers: A Cultural Anatomy of Women's Body Building.* New Brunswick, NJ: Rutgers University Press.

Hivon, M. 1994. Vodka: The "Spirit" of Exchange. *Cambridge Anthropology* 17: 1–18.

Hobsbawm, E.J. and T.O. Ranger. 1992. *The Invention of Tradition.* Cambridge: Cambridge University Press.

Hogden, M. 1974. *Anthropology, History and Cultural Change.* Viking Fund Publications in Anthropology, No. 52. Tuscon: University of Arizona Press for the Wenner-Gren Foundation for Anthropological Research.

Holton, M. 1974. *The Big Horse and Other Stories of Modern Macedonia.* Columbia: University of Missouri Press.

Holton, M. and G.W. Reid. 1977. *Reading the Ashes: An Anthology of the Poetry of Modern Macedonia.* Pittsburgh: University of Pittsburgh Press.

Holy, L. 1993. The End of Socialism in Czechoslovakia. In *Socialism: Ideals, Ideologies and Local Practice,* ed. C.M. Hann. London and New York: Routledge.

Honig, E. and G. Hershatter. 1988. *Personal Voices: Chinese Women in the 1980s.* Stanford: Stanford University Press.

Hubbard, R. 1990. *The Politics of Women's Biology.* Rutgers: Rutgers University Press.

Humphrey, C. 1995. Creating a Culture of Disillusionment: Consumption in Moscow, A Chronicle of Changing Times. In *Worlds Apart: Modernity Through the Prism of the Local,* ed. D. Miller, 43–68. London: Routledge.

Irigaray, L. 1991. *Ethik Der Sexuellen Differenz.* Frankfurt/Main: Suhrkamp Verlag.

Ivekovic, R. 1995. Women, Democracy and Nationalism After 1989: The Yugoslav Case. *Canadian Women Studies: Women in Central and Eastern Europe* 16: 10–13.

Jaggar, A.M. and S. Bordo. 1989. *Gender/Body/Knowledge: Feminist Reconstructions of Being and Knowing.* New Brunswick, NJ: Rutgers University Press.

Jancar, B.W. 1978. *Women Under Communism.* Baltimore: Johns Hopkins University Press.

Jancar, B.W. 1990. *Women and Revolution in Yugoslavia, 1941–45.* Women and Modern Revolution Series. Denver, CO: Arden Press.

Jelavich, B. 1984. *History of the Balkans: The Nineteenth and Twentieth Centuries.* Cambridge: Cambridge University Press.

Jenkins, C. and B. Sherman. 1979. *The Collapse of Work.* London: Eyre Methuen.

Jenkins, C. and B. Sherman. 1981. *The Leisure Shock.* London: Eyre Methuen.

Johnson, M. 1998. At Home and Abroad: Inalienable Wealth, Personal Consumption and the Formulations of Femininity in the Southern Philippines. In *Material Cultures: Why Some Things Matter*, ed. D. Miller, 215–40. Chicago: University of Chicago Press.

Jones, L. 1990. *States of Change: A Central European Diary Autumn, 1989*. London: Merlin.

Joyce, P. 1989. *The Historical Meanings of Work*. Cambridge: Cambridge University Press.

Kaldor, M. 1993. Yugoslavia and the New Nationalism. *New Left Review* 147: 96–112.

Karakasidou, A. 1993. Politicizing Culture: Negating Ethnic Identity in Greek Macedonia. *Journal of Modern Greek Studies* 11: 1–28.

Karakasidou, A.N. 1997a. *Fields of Wheat, Hills of Blood: Passages to Nationhood in Greek Macedonia, 1870–1990*. Chicago: University of Chicago Press.

Karakasidou, A.N. 1997b. Women of the Family, Women of the Nation: National Enculturation Among Slav-Speakers in North-West Greece. In *Ourselves and Others: The Development of Greek Macedonian Cultural Identity Since 1912*, ed. P. Mackridge and E. Yannakakis, 91–110. Oxford: Berg.

Karaosmanoglu, A.L. 1993. *Crisis in the Balkans*. New York: United Nations.

Katz, C. and J. Monk, eds. 1993. *Full Circles: Geographies of Women Over the Life Course*. London: Routledge.

Kendall, L. 1985. *Shamans, Housewives, and Other Restless Spirits: Women in Korean Ritual Life*. Studies of the East Asian Institute. Honolulu: University of Hawaii Press.

Kendall, L. 1988. *The Life and Hard Times of a Korean Shaman: Of Tales and the Telling of Tales*. Honolulu: University of Hawaii Press.

Kendall, L. 1996. *Getting Married in Korea: Of Gender, Morality, and Modernity*. Berkeley: University of California Press.

Kideckel, D. 1995. *East European Communities: The Struggle for Balance in Turbulent Times*. Boulder, CO: London: Westview Press.

Kitromilides, P. 1989. "Imagined Communities" and the Origins of the National Question in the Balkans. *European History Quarterly* 19: 149–94.

Kligman, G. 1988. *The Wedding of the Dead: Ritual, Poetics and Popular Culture in Transylvania*. Berkeley: University of California Press.

Koch, H.-G. 1990. German Reunification and the Abortion Law. *Planned Parenthood in Europe* 19: 4.

Kofos, E. 1964. *Nationalism and Communism in Macedonia*. Thessaloniki: Institute for Balkan Studies.

Kofos, E. 1987. *The Macedonian Question: The Politics of Mutation*. Thessaloniki: Institute for Balkan Studies.

Kofos, E. 1989. National Heritage and National Identity in Nineteenth- and Twentieth-Century Macedonia. *European History Quarterly* 19: 229–67.

Kondo, D.K. 1990. *Crafting Selves: Power, Gender, and Discourses of Identity in a Japanese Workplace.* Chicago: University of Chicago Press.

Korade, D. 1994. Struggling Against the Nationalist Right. *Balkan War Report* 30: 36.

Korobar, P. and O. Ivanoski. 1983. *Documents, Studies, Resolutions, Appeals and Published Articles. The Historical Truth: The Progressive Social Circles in Bulgaria and Pirin Macedonia on the Macedonian National Question 1896–1956.* Trans. Filip Korzenski. Skopje: Kultura.

Kottak, C.P. 1990. *Prime-Time Society: An Anthropological Analysis of Television and Culture.* Belmonte, CA: Wadsworth.

Kryshtanovskaya, O. 1992. The New Business Elite. In *Russia in Flux: The Political and Social Consequences of Reform,* ed. D. Lane, 185–95. London: Edward Elgar.

Kuper, A. and European Association of Social Anthropologists. 1992. *Conceptualizing Society.* London: Routledge.

Lambek, M.J. and A. Strathern. 1998. *Bodies and Persons: Comparative Perspectives From Africa and Melanesia.* Cambridge: Cambridge University Press.

Laqueur, T. 1991. *Making Sex: Body and Gender From the Greeks to Freud.* Cambridge, MA: Harvard University Press.

Latitic, I., ed. 1975. *Women in the Economy and Society of the SFR of Yugoslavia.* Belgrade: Federal Institute for Statistics.

Layoun, M. 1992. Telling Spaces: Palestinian Women and the Engendering of National Narratives. In *Nationalism and Sexualities,* ed. A. Parker, M. Russo, D. Sommer and P. Yaeger. London: Routledge.

Leach, E.R. 1990. *Political Systems of Highland Burma: A Study of Kachin Social Structure.* London: Athlone Press.

Leacock, E. 1981. *Myths of Male Dominance.* New York: Monthly Review Press.

Lele, J. 1981. *Tradition and Modernity in Bhakti Movements.* International Studies in Sociology and Social Anthropology, Vol. 31. Leiden: Brill.

Lemel, H. 2000. *Rural Property and Economy in Post-communist Albania.* New York; Oxford: Berghahn Books.

Levi-Strauss, C. 1969. *The Elementary Structures of Kinship.* London: Eyre and Spottiswoode.

Libal, W. 1993. *Mazedonien Zwischen Den Fronten: Junger Staat Mit Alten Konflikten.* Wien: Europaverlag.

Lindstrom, L. 1993. *Cargo Cult: Strange Stories of Desire From Melanesia and Beyond.* Honolulu: University of Hawaii Press.

Lockwood, W.G. 1975. *European Moslems: Economy and Ethnicity in Western Bosnia.* New York: Academic Press.

Loizos, P. 1975a. *The Greek Gift: Politics in a Cypriot Village.* Pavilion Series. Oxford: Blackwell.

Loizos, P. 1975b. Changes in Property Transfer among Greek Cypriot Villagers. *Man* 10: 503–23.

Loizos, P. 1981. *The Heart Grown Bitter: A Chronicle of Cypriot War Refugees.* Cambridge: Cambridge University Press.

Loizos, P. 1991. Gender and Kinship in Marriage and Alternative Contexts. In *Contested Identities: Gender and Kinship in Modern Greece,* ed. P. Loizos and E. Papataxiarchis, 3–25. Princeton: Princeton University Press.

Loizos, P. and E. Papataxiarchis, eds. 1991. *Contested Identities: Gender and Kinship in Modern Greece.* Princeton: Princeton University Press.

Loizos, P. and E. Papataxiarchis. 1991. Gender, Sexuality, and the Person in Greek Culture. In *Contested Identities: Gender and Kinship in Modern Greece,* ed. P. Loizos and E. Papataxiarchis, 221–35. Princeton: Princeton University Press.

Lunt, H. 1984. Some Sociolinguistic Aspects of Macedonian and Bulgarian. In *Language and Literary Theory: In Honor of Ladislav Matejka,* ed. L. Dolezel, I.R. Titunik and B.A. Stolz. Papers in Slavic Philology, No. 5. Ann Arbor: University of Michigan Slavic Publications.

Maccormack, C.P. 1982. *Ethnography of Fertility and Birth.* London: Academic Press.

Maccormack, C.P. and M. Strathern. 1980. *Nature, Culture, and Gender.* New York: Cambridge University Press.

Macedonia, G.O.T.R.O. 1995. *Women: Status of Equality. National Report of the Republic of Macedonia for the 4th World Conference on Women in Beijing, 1995.* Government of the Republic of Macedonia.

Macfarlane, A. 1986. *Marriage and Love in England: Modes of Reproduction 1300–1840.* Oxford: Basil Blackwell.

Madan, T.N.. 1978. *Dialectic of Tradition and Modernity in the Sociology of D.P. Mukerji: A Memorial Lecture Delivered on 25th February 1977 at the University of Lucknow.* D.P. Mukerji Memorial Lecture Series, No. 1. New Delhi: Manohar Publications for the University of Lucknow.

Mamonova, T., ed. 1984. *Women and Russia.* Feminist Writings From the Soviet Union. Boston: Beacon Press.

Marris, P. 1974. *Loss and Change.* New York: Pantheon Books.

Martin, E. 1987. *The Woman in the Body: A Cultural Analysis of Reproduction.* Boston: Beacon Press.

Martis, N.K. 1983. *The Falsification of Macedonian History.* Athens: Graphic Arts

Mauss, M. 1979. Body Techniques. In Sociology and Psychology, 95–123. London: Routledge.

Mauss, M. 1989. *The Gift: The Form and Reason for Exchange in Archaic Societies.* London: Routledge.

McAuley, A. 1992. Poverty and Underprivileged Groups. In *Russia in Flux: The Political and Social Consequences of Reform,* ed. D. Lane, 196–209. London: Edward Elgar.

McDonald, M. 1990. *"We Are Not French!" Language, Culture, and Identity in Brittany.* London: Routledge.

McDonogh, G.W. and R. Rotenberg. 1993. *The Cultural Meaning of Urban Space.* Contemporary Urban Studies. Westport, CT: Bergin and Garvey.

Meigs, A. 1984. *Food, Sex and Pollution: A New Guinea Religion.* New Brunswick: Rutgers University Press.

Mendras, H. and A. Cole. 1991. *Social Change in Modern France: Towards a Cultural Anthropology of the Fifth Republic.* Cambridge: Cambridge University Press.

Michie, J. and J. Grieve Smith. 1997. *Employment and Economic Performance: Jobs, Inflation, and Growth.* New York: Oxford University Press.

Miller, B., ed. 1992. *Sex and Gender Hierarchies.* Cambridge: Cambridge University Press.

Miller, D. 1987. *Material Culture and Mass Consumption.* Oxford: Blackwell.

Miller, D. 1993. *Unwrapping Christmas.* Oxford: Clarendon Press.

Miller, D. 1994. *Modernity, an Ethnographic Approach: Dualism and Mass Consumption in Trinidad.* Oxford: Berg.

Miller, D. 1997. *Capitalism: An Ethnographic Approach.* Explorations in Anthropology. Oxford: Berg.

Miller, D. 1998. *Material Cultures: Why Some Things Matter.* Chicago: University of Chicago Press.

Miller, D. 1998. *A Theory of Shopping.* Cambridge: Polity Press in Association with Blackwell.

Mills, C.W. 1951. *White Collar: The American Middle Class.* New York: Oxford University Press.

Moore, H.L. 1988. *Feminism and Anthropology.* Cambridge: Polity Press.

Moore, H.L. 1993. The Differences Within and the Differences Between. In *Gendered Anthropology,* ed. T.D. Valle, 193–204. London: Routledge.

Moore, H.L. 1994. *A Passion for Difference: Essays in Anthropology and Gender.* Cambridge: Polity Press.

Morokvasic, M. 1986. Being a Woman in Yugoslavia: Past, Present and Institutional Equality. In *Women of the Mediterranean,* ed. M. Gadant, 120–38. London: Zed Books.

Mosely, P.E. 1976. The Zadruga. In *Communal Families in the Balkans,* ed. R.F. Byrnes. Notre Dame: University of Notre Dame Press.

Nash, J.C., J.E. Corradi and H. Spalding. 1977. *Ideology and Social Change in Latin America.* New York: Gordon and Breach.

Nash, J.C. and M.P. Fernández-Kelly. 1983. *Women, Men, and the International Division of Labor.* SUNY Series in the Anthropology of Work. Albany: State University of New York Press.

Nodia, G. 1993. Nationhood and Self-Recollection: Ways to Democracy After Communism. In *Towards a New Community: Culture and Politics in Post-Totalitarian Europe,* ed. P.J.S. Duncan and M. Rady, 53–64. Hamburg and Münster: LIT Verlag.

Noller, P. and V. Callan. 1991. *The Adolescent in the Family.* Adolescence and Society. London: Routledge.

O'Hanlon, M. 1983. Handsome Is as Handsome Does: Display and Betrayal in the Waghi. *Oceania* 56: 181–98.

O'Hanlon, M. and L. Frankland. 1986. With a Skull in the Netbag: Prescriptive Marriage and Matrilateral Relations in the New Guinea Highlands. *Oceania* 56: 181–98.

Oakes, G. 1989. Four Questions Concerning the Protestant Ethic. *Telos* 81: 77–85.

Oakley, A. 1972. *Sex, Gender and Society.* London: Temple Smith.

Office of Statistics. 1991. "Dame Gruev" 4–Skopje, Republic of Macedonia.

Okely, J. 1983. *The Traveller Gypsies.* London: Cambridge University Press.

Ong, A. 1987. *Spirits of Resistance and Capitalist Discipline: Factory Women in Malaysia.* SUNY Series in the Anthropology of Work. Albany: State University of New York Press.

Ong, A. and M.G. Peletz. 1995. *Bewitching Women, Pious Men: Gender and Body Politics in Southeast Asia.* Berkeley: University of California Press.

Orbach, S. 1978. *Fat Is a Feminist Issue: The Anti-Diet Guide to Permanent Weight Loss Susie Orbach.* New York: Berkley Books.

Orbach, S. 1986. *Hunger Strike: An Anorectic's Struggle as a Metaphor for Our Age.* New York: Norton.

Ortner, S. 1974. Is Female to Male as Nature Is to Culture? In *Women, Culture and Society*, ed. M.Z. Rosaldo and L. Lamphere, 67–87. Stanford: Stanford University Press.

Ortner, S. and H. Whitehead. 1981. Accounting for Sexual Meanings. In *Sexual Meanings: The Cultural Construction of Gender and Sexuality*, ed. S. Ortner and H. Whitehead, 1–28. Cambridge: Cambridge University Press.

Ortner, S. and H. Whitehead, eds. 1981. *Sexual Meanings: The Cultural Construction of Gender and Sexuality.* Cambridge: Cambridge University Press.

Ossman, S. 1994. *Picturing Casablanca: Portraits of Power in a Modern City.* London: University of California Press.

Overing, J. 1986. Men Control Women? The 'Catch 22' in the Analysis of Gender. *International Journal of Moral and Social Studies* 1 (2): 135–56.

Pahl, R.E., ed. 1988. *On Work: Historical, Comparative and Theoretical Approaches.* Oxford: Basil Blackwell.

Palmer, S.E. and R.R. King. 1971. *Yugoslav Communism and the Macedonian Question.* Hamden, CT: Archon Books.

Parker, A., M. Russo, D. Sommer and P. Yaeger, eds. 1992. *Nationalisms and Sexualities.* London: Routledge.

Pawlik, W. 1992. Intimate Commerce. In *The Unplanned Society: Poland During and After Communism*, ed. J. Wedel, 78–94. New York: Columbia University Press.

Peacock, J.L. 1968. *Rites of Modernization; Symbolic and Social Aspects of Indonesian Proletarian Drama.* Chicago: University of Chicago Press.

Pells, R.H. 1997. *Not Like Us: How Europeans Have Loved, Hated, and Transformed American Culture Since World War II.* New York: Basic Books.

Pemberton, J. 1994. *On the Subject of "Java."* Ithaca: Cornell University Press.

Perry, D.M. 1988. *The Politics of Terror: The Macedonian Liberation Movements, 1893–1903.* Durham: Duke University Press.

Pesic, V. 1991. The Impact of Reforms on the Status of Women in Yugoslavia. Paper Presented to the Regional Seminar on the Status of Women in Eastern Europe and the Soviet Union, Vienna.

Petersen, A. 1985. *Ehre Und Scham: Das Verhältnis Der Geschlechter In Der Türkei.* Berlin: Express-Edition.

Phizacklea, A., H. Pilkington and S. Rai, eds. 1992. *Women in the Face of Change.* London: Routledge.

Pina-Cabral, J.D. 1986. *Sons of Adam, Daughters of Eve: The Peasant Worldview of the Alto Minho.* Oxford: Clarendon Press.

Pine, F. 1987. Kinship, Marriage and Social Change in a Polish Highland Village. Ph.D. diss., University of London.

Pine, F. 1993. "The Cows and Pigs Are His, The Eggs Are Mine": Women's Domestic Economy and Entrepreneurial Activity in Rural Poland. In *Socialism: Ideals, Ideologies and Local Practice,* ed. C.M. Hann, 227–24. London: Routledge.

Pine, F. 1994. Privatisation in Post-Socialist Poland: Peasant Women, Work, and the Restructuring of the Public Sphere. *Cambridge Anthropology* 17: 19–42.

Pine, F. 1996. Redefining Women's Work in Rural Poland. In *After Socialism: Land Reform and Rural Social Change in Eastern Europe,* ed. R. Abrahams, 133–55. Oxford: Berghahn Books.

Pitkin, D.S. 1993. Italian Urban Space: Intersection of Private and Public. In *The Cultural Meaning of Urban Space,* ed. G.W. Mcdonogh and R. Rotenberg. Contemporary Urban Studies. Westport: Bergin and Garvey.

Pitt-Rivers, J. [1954] 1971. *The People of the Sierra.* Chicago: University of Chicago Press.

Portuges, C. 1992. Lovers and Workers: Screening the Body in Post-Communist Hungarian Cinema. In *Nationalisms and Sexualities,* ed. A. Parker, M. Russo, D. Sommer and P. Yaeger, 296–312. London: Routledge.

Poulton, H. 1995. *Who Are the Macedonians?* Bloomington: Indiana University Press.

Powdermaker, H. 1950. *Hollywood the Dream Factory: An Anthropologist Looks at the Movie Makers.* Boston: Little, Brown.

Powers, M.N. 1986. *Oglala Women: Myth, Ritual, and Reality.* Chicago: University of Chicago Press.

Pribichevich, S. 1982. *Macedonia: Its People and History.* University Park, PA: Pennsylvania State University Press.

Radhakrishnan, R. 1992. Nationalism, Gender, and the Narrative of Identity. In *Nationalisms and Sexualities*, ed. A. Parker, M. Russo, D. Sommer and P. Yaeger, 77–95. London: Routledge.

Rafael, V.L. 1995. *Discrepant Histories: Translocal Essays on Filipino Cultures.* Asian American History and Culture Series. Philadelphia: Temple University Press.

Ramet, S.P. 1984. Women, Work and Self-Management in Yugoslavia. *East European Quarterly* 17 (4): 459–68.

Ramet, S.P. 1985. Apocalypse Culture and Social Change in Yugoslavia. In *Yugoslavia in the 1980s*, ed. S.P. Ramet, 3–26. Colorado: Westview Press.

Ramet, S.P. 1995. *Social Currents in Eastern Europe: The Sources and Consequences of the Great Transformation.* London: Duke University Press.

Randal, V. 1982. *Women and Politics.* London: Macmillan Press.

Rausing, S. 1998. Signs of the New Nation: Gift Exchange, Consumption and Aid on a Former Collective Farm in North-West Estonia. In *Material Cultures: Why Some Things Matter*, ed. D. Miller, 189–213. Chicago: University of Chicago Press.

Reading, A. 1992. *Polish Women, Solidarity and Feminism.* London: Macmillan.

Reiter, N.N., ed. 1987. *Die Stellung Der Frau Auf Dem Balkan: Beitrag Zur Tagung Vom 3.-7. September 1985 In Berlin.* Balkanologische Veröffentlichungen, Bd. 12. Wiesbaden: Harrassowitz.

Reiter, R.R., ed. 1975. *Toward an Anthropology of Women.* New York: Monthly Review Press.

Rheubottom, D. 1971. A Structural Analysis of Conflict and Cleavage in Macedonian Domestic Groups. Ph.D. diss., University of Rochester.

Rheubottom, D. 1976a. The Saint's Feast and Skopska Crna Gora Social Structure. *Man* 11: 18–34.

Rheubottom, D. 1976b. Time and Form: Contemporary Macedonian Households and the Zadruga Controversy. In *Communal Families in the Balkans*, ed. R. Byrnes. Notre Dame: University of Notre Dame Press.

Rheubottom, D. 1980. Dowry and Wedding Celebrations in Yugoslav Macedonia. In *The Meaning of Marriage Payments*, ed. J. Comaroff, 221–49. London: Academic Press.

Rheubottom, D. [1985] 1993. The Seed of Evil Within. In *The Anthropology of Evil*, ed. D. Parkin, 12–24. Cambridge: Blackwell.

Rich, A. 1986. *Blood, Bread, and Poetry: Selected Prose, 1979–1985.* New York: Norton.

Roberts, K. 1970. *Leisure.* London: Longman.

Rosaldo, M.Z. and L. Lamphere, eds. 1974. *Women, Culture and Society.* Palo Alto: Stanford University Press.

Rubin, G. 1975. The Traffic in Women: Notes on the "Political Economy" of Sex. In *Toward an Anthropology of Women*, ed. R. Reiter, 157–88. New York: Monthly Review Press.

Rutherford, J. 1988. Who's That Man? In *Male Order: Unwrapping Masculinity*, ed. R. Chapman and J. Rutherford, 21–67. London: Lawrence and Wishart.

Samuel, R. and P. Thompson. 1990. *The Myth We Live By*. London: Routledge.

Sanday, P.R. 1981. *Female Power and Male Dominance: On the Origins of Sexual Inequality*. Cambridge: Cambridge University Press.

Sanday, P.R. and R. Goodenough, eds. 1990. *Beyond the Second Sex: New Directions in the Anthropology of Gender*. Philadelphia: University of Pennsylvania Press.

Sant Cassia, P. and C. Bada. 1992. *The Making of the Modern Greek Family: Marriage and Exchange in Nineteenth-Century Athens*. Cambridge Studies in Social and Cultural Anthropology. Cambridge: Cambridge University Press.

Sayers, J., M. Evans and N. Redclift, eds. 1987. *Engels Revisited: New Feminist Essays*. London: Tavistock.

Schneider, D. [1968] 1980. *American Kinship: A Cultural Account*. Chicago: University of Chicago Press.

Schor, J. 1991. *The Overworked American: The Unexpected Decline of Leisure*. New York: Basic Books.

Schor, N. and E. Weed. 1994. *More Gender Trouble: Feminism Meets Queer Theory*. Differences: A Journal of Feminist Cultural Studies, Vol. 6. Bloomington: Indiana University Press.

Schwalbe, M.L. 1988. Sources of Self-Esteem in Work. *Work and Occupations* 15: 24–35.

Schwimmer, E. 1984. The Self and the Product: Concepts of Work in Comparative Perspective. In *Social Anthropology of Work*, ed. S. Wallman, 287–315. London: Academic Press.

Scott, H. 1976. *Women and Socialism: Experiences From Eastern Europe*. London: Allison and Busby.

Scott, J.W. and L.A. Tilly. 1975. Women's Work and the Family in Nineteenth-Century Europe. *Comparative Studies in Society and History* 17: 36–64.

Segalen, M. [1980] 1983. *Love and Power in the Peasant Family: Rural France in the Nineteenth Century*. Trans. Sarah Matthews. Oxford: Basil Blackwell.

Segalen, M. [1981] 1986. *Historical Anthropology of the Family*. Trans. J.C. Whitehouse and Sarah Matthews. Cambridge: Cambridge University Press.

Sekelj, L. [1992] 1993. *Yugoslavia: The Process of Disintegration*. Trans. Vera Vukelic. Atlantic Studies on Societies in Change, No. 76. Boulder: Columbia University Press.

Shanin, T. 1988. *Peasants and Peasant Society*. Harmondsworth: Penguin.

Shields, R. 1992. *Lifestyle Shopping: The Subject of Consumption*. The International Library of Sociology. London: Routledge.

Simic, A. 1973a. Kinship Reciprocity and Rural-Urban Integration in Serbia. *Urban Anthropology* 2: 205–13.
Simic, A. 1973b. The Best of Two Worlds: Serbian Peasants in the City. In *Anthropologists in Cities*, ed. G.M. Foster and R.V. Kemper, 179–200. Boston: Little, Brown.
Simic, A. 1974. Urbanisation and Cultural Process in Yugoslavia. *Anthropological Quarterly* 47: 211–27.
Simic, A. 1983. Urbanization and Modernization in Yugoslavia. In *Urban Life in Mediterranean Europe: Anthropological Perspectives*, ed. M. Kenny and D. Kertzer, 203–25. Chicago: University of Illinois Press.
Simmel, G. 1978. *The Philosophy of Money.* London: Routledge and Kegan Paul.
Skelton, T. and G. Valentine. 1998. *Cool Places: Geographies of Youth Cultures.* London: Routledge.
Sorabji, C. 1989. Muslim Identity and Islamic Faith in Socialist Sarajevo. Ph.D. diss., University of Cambridge.
Southall, A. 1992. Marx, Engels, Lenin and the Anthropology of Change. In *The Curtain Rises: Rethinking Culture, Ideology and the State in Eastern Europe*, ed. M. Desoto and D.G. Anderson. New York: Humanities Press.
Spangler, M. 1983. Urban Research in Yugoslavia. In *Urban Life in Mediterranean Europe: Anthropological Perspectives*, ed. M. Kenny and D. Kertzer. Chicago: University of Illinois Press.
Spittler, G. 1991. Die Arbeitswelt In Agrargesellschaften. *Kölner Zeitschrift Für Soziologie Und Sozialpsychologie* 43: 1–17.
Spyer, P., ed. 1998. *Border Fetishisms: Material Objects in Unstable Spaces.* Zones of Religion. London: Routledge.
Stacey, J. 1983. *Patriarchy and Socialist Revolution in China.* Berkeley: University of California Press.
Stephens, S. 1995. *Children and the Politics of Culture.* Princeton Studies in Culture/Power/History. Princeton: Princeton University Press.
Stewart, C. 1991. *Demons and the Devil: Moral Imagination in Modern Greek Culture.* Princeton: Princeton University Press.
Stewart, M. 1987. Brothers in Song: The Persistence of (Vlach) Gypsy Identity and Community in Socialist Hungary. Ph.D. diss., University of London.
Stewart, M. 1993. Gypsies, the Work Ethic, and Hungarian Socialism. In *Socialism: Ideals, Ideologies and Local Practice*, ed. C.M. Hann, 187–203. London: Routledge.
Strathern, M. 1980. No Name, No Culture: The Hagen Case. In *Nature, Culture, and Gender*, ed. C. MacCormack and Marilyn Strathern, 174–222. Cambridge: Cambridge University Press.
Strathern, M. 1981a. Culture in a Netbag: The Manufacture of a Subdiscipline in Anthropology. *Man* 16: 665–88.

Strathern, M. 1981b. Self-Interest and the Social Good: Some Implications of Hagen Gender Imagery. In *Sexual Meanings: The Cultural Construction of Gender and Sexuality*, ed. S. Ortner and H. Whitehead, 166–91. Cambridge: Cambridge University Press.

Strathern, M. 1987a. An Awkward Relationship: The Case of Feminism and Anthropology. *Signs* 12: 276–92.

Strathern, M., ed. 1987b. *Dealing With Inequality: Analysing Gender Relations in Melanesia and Beyond*. Cambridge: Cambridge University Press.

Strathern, M. 1988. *The Gender of the Gift*. Berkeley: University of California Press.

Strathern, M. 1992. *After Nature: English Kinship in the Late Twentieth Century*. The Lewis Henry Morgan Lectures, 1984. Cambridge: Cambridge University Press.

Taussig, M. 1980. *The Devil and Commodity Fetishism in South America*. Chapel Hill: University of North Carolina Press.

Thiessen, I. 1999a. *T'Ga za Jug*. Ph. D London School of Economics.

Thiessen, I. 1999b. The Essence of Being: Procreation and Sexuality in Mid-Century Macedonia. In *Conceiving Persons: Ethnographies of Procreation, Fertility and Growth*, eds. P. Loizos and P. Heady, 177-99. London: Athlone Press.

Thomas, K. 1964. Work and Leisure in Pre Industrial Society. *Past and Present* 29: 50–66.

Thompson, E.P. 1967. Time, Work-Discipline, and Industrial Capitalism. *Past and Present* 38: 56–97.

Tilly, L. and J. Scott. 1978. *Women, Work and Family*. New York: Holt, Rinehart and Winston.

Todorova, M.N. 1992. *Balkan Family Structure and the European Pattern: Demographic Developments in Ottoman Bulgaria*. Washington: American University Press.

Todorova, M. 1997. *Imagining the Balkans*. Oxford: Oxford University Press.

Tonkin, E., M. McDonald and M. Chapman, eds. 1989. *History and Ethnicity*. London: Routledge.

Touraine, A. 1977. *The Self-Production of Society*. Trans. Derek Coltman. Chicago and London: University of Chicago Press.

Tsing, A.L. 1993. *In the Realm of the Diamond Queen: Marginality in an Out-Of-The-Way Place*. Princeton: Princeton University Press.

Turner, V.W. 1967. *The Forest of Symbols: Aspects of Ndembu Ritual*. Ithaca: Cornell University Press.

United States. Congress. Commission on Security and Cooperation in Europe. 1995. *The United Nations, NATO, and the Former Yugoslavia: Hearing Before the Commission on Security and Cooperation in Europe, One Hundred Fourth Congress, First Session, April 6, 1995*. Washington: U.S. G.P.O.

United States. President (1993–2001: Clinton).1994. *Follow-Up Report to the Committee of Foreign Affairs on Former Yugoslav Republic of Macedonia: Communication From the President of the United States Transmitting His Follow-Up Report on the United States Peacekeeping Contingent in the Former Yugoslav Republic of Macedonia.* Washington: U.S. G.P.O.

United States. President (1993–2001: Clinton).1994. *Further Reporting on U.S. Forces in the Republic of Macedonia to the Committee on Foreign Affairs: Communication From the President of the United States Transmitting His Further Report Concerning His Decision to Deploy a U.S. Army Peacekeeping Contingent as Part of the United Nations Protection Force in the Republic of Macedonia.* Washington: U.S. G.P.O.

United States. President (1993–2001: Clinton).1995. *Status Report to the Committee on International Relations on the Former Yugoslav Republic of Macedonia: Communication From the President of the United States Transmitting His Fourth Report on the Continuing Deployment of a U.S. Army Peacekeeping Contingent as Part of the United Nations Protection Force (UNPROFOR) in the Former Yugoslav Republic of Macedonia (FYROM), Consistent With the War Powers Resolution.* Washington: U.S. G.P.O.

Uusitalo, L. 1979. *Consumption Style and Way of Life.* Helsinki: Helsinki School of Economics.

Vasery, I. 1987. *Beyond the Plan: Social Change in a Hungarian Village.* Boulder: Westview Press.

Vasiliadis, P. 1989. *Whose Are You?: Identity and Ethnicity Among the Toronto Macedonians.* New York: AMS Press.

Veal, A.J. 1987. *Leisure and the Future.* London: Allen and Unwin.

Veal, A.J. 1989. Leisure and the Future: Considering the Options. In *Freedom and Constraints,* ed. F. Coalter. New York: Routledge.

Verdery, K. 1991. Theorizing Socialism: A Prologue to the "Transition." *American Ethnologist* 18: 419–39.

Verdery, K. 1993. Ethnic Relations, Economies of Shortage, and the Transition in Eastern Europe. In *Socialism: Ideals, Ideologies and Local Practice,* ed. C.M. Hann, 172–86. London and New York: Routledge.

Verdery, K. 1995. "Caritas" and the Reconceptualization of Money in Romania. *Anthropology Today* 11: 3–7.

Veremis, T. and M. Blinkhor, eds. [1989] 1990. *Modern Greece: Nationalism and Nationality.* Athens: Sage-Eliamep.

Vogel, L. 1983. *Marxism and the Oppression of Women.* London: Pluto Press.

Wadel, C. 1984. The Hidden Work of Everyday Life. In *Social Anthropology of Work,* ed. S. Wallman, 165–84. London: Academic Press.

Wallman, S. 1992. *Contemporary Futures: Perspectives From Social Anthropology.* ASA Monographs, No. 30. London: Routledge.

Wallman, S. 1979. *Social Anthropology of Work.* A.S.A. Monograph, No. 19. New York: Academic Press.

Walzer, M. 1992. *Just and Unjust Wars: A Moral Argument with Historial Illustrations.* New York: Basic Books.

Watson, R. 1984. Women's Property in Republican China: Rights and Practice. *Republican China* 10: 1–12.

Watson, R.S., ed. 1994. *Memory, History, and Opposition Under State Socialism.* Santa Fe: School of American Research Press.

Weber, M. 1958. *The Protestant Ethic and the Spirit of Capitalism.* Trans. Talcott Parson. New York: Charles Scribner's Sons.

Wedel, J. 1986. *The Private Poland.* New York: Facts on File Inc.

Wedel, J., ed. 1992. *The Unplanned Society: Poland During and After Communism.* New York: Columbia University Press.

Weigand, G. 1924. *Ethnographie Von Makedonien.* Leipzig: Friedrich Brandstetter.

Wheeler, M. 1993. Reconstruction as Deconstruction: The Case of Yugoslavia. In *Towards a New Community: Culture and Politics in Post-Totalitarian Europe,* ed. P.J.S. Duncan and M. Rady, 121–32. Hamburg and Münster: LIT Verlag.

Whitehead, H. 1981. The Bow and the Burden Strap: A New Look at Institutionalized Homosexuality in Native North America. In *Sexual Meanings: The Cultural Construction of Gender and Sexuality,* ed. S. Ortner and H. Whitehead, 80–115. Cambridge: Cambridge University Press.

Whitehead, T.L. and M.E. Conaway, eds. 1986 *Self, Sex and Gender in Cross-Cultural Fieldwork.* Urbana: University of Illinois Press.

Wikan, U. 1977. Man Becomes Woman: Transsexualism in Oman as a Key to Gender Roles. *Man* 12: 304–19.

Wikan, U. 1991. *Behind the Veil in Arabia: Women in Oman.* Chicago: University of Chicago Press.

Wilkinson, H.R. 1951. *Maps and Politics: A Review of the Ethnographic Cartography of Macedonia.* Liverpool: Liverpool University Press.

Wirth, L. 1928. *The Ghetto.* Chicago: University of Chicago Press.

Wolf, E.R., A. Koster and D. Meijers. 1991. *Religious Regimes and State-Formation: Perspectives From European Ethnology.* Albany: State University of New York Press.

Wolf, N. 1992. *The Beauty Myth: How Images of Beauty Are Used Against Women.* New York: Doubleday.

Woodward, K. 1997. *Identity and Difference.* Culture, Media and Identities, Vol. 3. London: Sage in Association with the Open University.

Worsley, P. 1968. *The Trumpet Shall Sound: A Study of "Cargo" Cults in Melanesia.* London: Macgibbon and Kee.

Wright, S. 1993. "Working Class" Versus "Ordinary People": Contested Ideas of Local Socialism in England. In *Socialism: Ideals, Ideologies and Local Practice,* ed. C.M. Hann. London and New York: Routledge.

Yaeger, P., ed. 1996. *The Geography of Identity.* Ann Arbor: University of Michigan Press.

Yanagisako, S.J. 1985. *Transforming the Past: Tradition and Kinship Among Japanese Americans.* Stanford: Stanford University Press.

Yanagisako, S.J. 1987. Mixed Metaphors: Native and Anthropological Models of Gender and Kinship Domains. In *Gender and Kinship: Essays Toward a Unified Analysis,* ed. J.F. Collier and S.J. Yanagisako, 86–118. Stanford: Stanford University Press.

Yedlin, T. 1978. Women in Eastern Europe and the Soviet Union. Paper Presented to the Conference on Women in Eastern Europe and the Soviet Union, University of Alberta. Division of East European Studies.

Yugoslavia. Sekretarijat Za Informacije. 1963. *This Was Skopje.* Beograd: Federal Secretariat for Information.

Zarkov, D. 1997. Sex as Usual: Body Politics and the Media War in Serbia. In *Embodied Practices: Feminist Perspectives on the Body,* ed. K. Davis, 110–30. Thousand Oaks, CA: Sage.

Zografski, D. 1990. Social-Development of Yugoslavian States and Especially Macedonia After the First-World-War. *East European Quarterly* 24: 237–46.

Index